Why Toast Lands Jelly-Side Down

Some other recent books by the author

Turning the World Inside Out, 1990

The Cosmological Milkshake, 1993

Electricity and Magnetism Simulations(CUPS), 1995

What If You Could Unscramble an Egg?, 1996

Why Toast Lands Jelly-Side Down

ZEN AND THE ART OF PHYSICS DEMONSTRATIONS

Robert Ehrlich

PRINCETON UNIVERSITY PRESS · PRINCETON, NEW JERSEY

Copyright © 1997 by Princeton University Press

Published by Princeton University Press, 41 William Street, Princeton, New Jersey 08540

In the United Kingdom: Princeton University Press, Chichester, West Sussex

Library of Congress Cataloging-in-Publication Data

Ehrlich, Robert, 1938-
Why toast lands jelly-side down : zen and the art of physics demonstrations / Robert Ehrlich.
p.   cm.
Includes bibliographical references and index.
ISBN 0-691-02891-5 (cloth : alk. paper). – ISBN 0-691-02887-7 (pbk. : alk. paper)
1. Physics–Experiments.   I. Title.
QC33.E55   1997
530'.078–dc20   96-42053 CIP

This book has been composed in Times Roman

Princeton University Press books are printed on acid-free paper and meet the guidelines for permanence and durability of the Committee on Production Guidelines for Book Longevity of the Council on Library Resources

10 9 8 7 6 5 4 3 2 1

10 9 8 7 6 5 4 3 2 1
(Pbk.)

This book is dedicated

to the memory of my mother-in-law,

Ruth Levy

Contents

viii

Contents

Contents

x

Contents

Acknowledgments

I am grateful to the following people who gave me feedback on an earlier draft of this book: Ron Edge, Peter Hopkinson, John Hubisz, Diandra Leslie-Pelecky, Bruce Sherwood, and Peter Volkovitsky.

Why Toast Lands Jelly-Side Down

Chapter 1

How to Design Simple Physics Demos

1.1 Introduction

This is not another one of those book combining Eastern philosophy and the ideas of modern physics. Rather, it is a practical book of physics demonstrations and experiments, and it includes an introductory chapter on how to design your own demonstrations. The "Zen" in the subtitle reflects my belief that the best demos are simple ones based on everyday objects and phenomena. Readers familiar with my earlier book, *Turning the World Inside Out and 174 Other Simple Physics Demonstrations*,[1] might consider this book to be a sequel, but the books have almost no overlap in terms of the listed demonstrations.

I expect that many readers of this book will be physics or physical science teachers at various educational levels, all the way to the university level. For such readers, this book could be used both as a resource book on how to design demos, and as a source of many low-cost demos for use in physics courses or exhibits. In addition it could serve as a text for a hands-on "advanced elementary physics" course—either formal or self-study. However, I have not assumed that all readers will be teachers, and in particular, I have not assumed that the group you wish to show the demos to is necessarily a physics class. Therefore, throughout the book I have used such generic terms "audience" and "observers" for the group to which you will be showing the demos. In addition to teachers, the book should also be of interest to students and others who enjoy seeing how the principles of physics relate to everyday objects, and how these principles can explain the world around them. In fact, most of the demos are sufficiently simple that they can be done using materials found around the home or office.

Unlike most existing physics demo manuals, which assume the basic principles underlying each demo are well understood, this book discusses these principles within the context of each demonstration, but it does presuppose a basic understanding of physics. While all the demos are simple, in terms of the materials they use, some are extremely challenging in terms of the physics concepts they involve. On the other hand, I have made a conscious effort to make the demos as suitable as possible for a wide range of educational levels, and both qualitative and quantitative parts of most demos have been included.

No claim is made that most of the demos discussed in this book are new, although many are, and many others involve adding new elements to well-known demos. Wherever possible, I have cited the source of demos that have been adopted from others, but in many cases, the cited source probably would not claim to be the originator. The process of designing your own demos is much like the process of scientific research itself, and it often involves starting with existing ideas of others, and modifying them in various ways.

1.2 Designing your own simple demos

An obvious first question about designing your own demos is "why bother?" A somewhat different answer to this question can be given to those readers who happen to be teachers, and those who happen to be present or former physics students. For those of you who are teachers at schools lacking adequate equipment budgets, part of the reason for designing your own demos should be obvious. Many homegrown versions of demos are considerably less expensive than their cousins found in the catalogs of the scientific equipment companies, and they do the job just as well. Moreover, by designing your own demos, you may actually be able to improve on existing ones, and customize them for your own needs.

Some people believe that students are turned off by overly simple apparatus that seems to lack a degree of scientific credibility. But I believe that just the opposite is true. Physics is often perceived as being highly abstract, user-unfriendly, and remote from everyday life. Using "low tech" items that you can work with at home makes physics more relevant to

**How to Design Simple
Physics Demos**

everyday life, and it humanizes it—while doing experiments with complex apparatus may have just the opposite effect. Obviously, I don't argue against using complex apparatus when necessary, only avoiding it if possible.

One important benefit of designing your own demos is that you get to understand them better— both in the sense of becoming more aware of their limitations and quirks, but more important, in terms of the basic physics behind the demo. Much of elementary physics is not so elementary, and even experienced physicists get surprised about the outcome of some "simple" demonstrations more often than they might care to admit. Perhaps the most important reason for teachers to design their demos is that they will be more likely to use them in their teaching.

Most physics teachers love to see a good demo, but they rarely show them in their classes—it is just too much of a bother. Who wants to bother tracking down the equipment you need, making sure it is set up, and possibly having to explain to the class what really should have happened when the demo fails to work! On the other hand, most of the simple home-grown demos discussed in this book have none of these drawbacks. You can fit many of them in your pocket, bring them to class without any set up required, and best of all, you need not fear that your demo will more likely illustrate Murphy's laws rather than Newton's.

Finally, you may want to develop your own demos as a way of developing your creative impulse, and sharing your ideas with the rest of the world—either in the sense of putting on a "physics show," creating a hands-on science exhibit, or writing about the demos you may develop. In any case, do be sure that the demo works as advertised. There are any number of demos that should work in principle, but have very serious practical difficulties, which are sometimes not obvious until one has spent a considerable amount of time on them. I confess to being a "sinner" in this regard, having included in some past publications a demo or two that I didn't actually try myself, but which I naively thought couldn't possibly fail to work. There can be no doubt that there is no sub-

stitute to doing an experiment yourself—experience
counts!

1.3 Places to get ideas for designing demos

You can get good ideas for demos in many places.
Obvious sources would include lab manuals, the
demo manuals listed in the bibliography, and items
listed in the catalogs of scientific equipment companies or odd-lot and surplus companies. Also,
don't forget to consult the journals devoted to physics teaching, including The Physics Teacher, The
American Journal of Physics, as well as two European journals—Physics Education, and the European
Journal of Physics. The American Journal of Physics
is especially good for more advanced topics likely to
be of interest to instructors at the university level,
while The Physics Teacher covers a wider spectrum
of levels. In addition, don't forget to read other journals dealing with science education generally, such as
the Journal of College Science Teaching, The Science
Teacher, and popular magazines such as Scientific
American.

Other good places to get good demo ideas would
include the national and regional meetings of the
American Association of Physics Teachers, and other
such organizations. Those who have access to the
World Wide Web can also find physics demo collections by doing a key word search. At the time
of this writing, the Physics Instructional Resources
Association (PIRA) maintains a website containing links to demonstration collections maintained by
nine universities.[2] My favorite places to get good
ideas for demos are various types of stores, such
as stores selling toys, novelties, crafts, hardware,
electronics, office supplies, and even supermarkets.
Specialized stores in science museums, public TV
stores, magic shops, and science or nature stores are
also very good places to get ideas or purchase existing demos. But, in scouring the world for good demo
ideas, don't forget that lots of good demo ideas may
crop up as close to home as your own classes—
particularly in the form of questions raised by
students.

1.4 Starting points in designing a "new" demo

The most obvious starting point for a new demo is an existing demo that you seek to improve or expand to a new domain in some way. The improvement might involve making it easier, more reliable, more portable, etc., or it might involve the way the demo is presented and analyzed. Expansion to a new domain could include, for example, using a method for observing interference for one type of waves and applying it to waves of another type. Given a goal of designing simple demos, an obvious starting point would be to change a measuring technique so that it could be accomplished using much simpler apparatus, for example, using a folded index card as a "force probe"—see demonstration 2.3.

Other starting points for a new demo might include either a particular physical principle you wish to illustrate, or perhaps some natural phenomenon you wanted to model, such as a sunset or a mirage. My most fruitful ideas for new demos come from the capabilities of some physical object—either a very ordinary object such as a plastic ruler or a ballpoint pen, or a special purpose toy or gadget. But, getting such good ideas more often comes as a result of actually playing with the object, rather than simply thinking about it. Three of the demonstrations in this book using magnetic marbles (3.8, 4.4, and 12.2), did not occur to me until I actually played with this amusing toy. Sometimes a simple object can have a wondrous range of uses in many demos. For example, a piece of folded index card—used as the equivalent of a gentle spring when it is allowed to unfold—not only makes an excellent force probe, but it can also serve to launch balls with equal and opposite momenta (demo 5.1), serve as a potential barrier to be overcome in a simulation of the fusion reaction (demo 12.2), or a means of reflecting rolling balls back to their point of origin (demo 12.8).

Finally, one fruitful way of getting good ideas for demos involves thinking about the capabilities of the overhead projector (OHP), which can make a small demo visible to an entire group—and thereby allow you to design small, compact, and low-cost versions of existing demos. But, the OHP has many less obvious additional benefits. It can serve as a source of heat as well as light. You can make the projector surface slightly inclined, and show slow-motion parabolic trajectories of a ball—in effect having "diluted" gravity. You can turn the projector on its side to give shadow projections. Or, you can change the focus to view any plane through a partly transparent object.

You may find that there is a considerable distance between getting what seems to be a good idea for a demo, and being able to convert the idea into a simple demo that works reliably. Actually trying out your ideas usually will prove a surer road to progress than simply thinking about them. Physical objects— even simple ones—sometimes behave in unexpected ways that may make a demo idea either more or less feasible than you originally thought.

In order to design good simple demos, however, it is important to keep in mind a set of criteria that good demos should satisfy. Once you have such a set of criteria in mind, the process of refining your demo ideas can then be expressed in terms of the following procedure:

- List all the criteria of good design your demo fails to satisfy.
- Think about how to modify the demo to satisfy as many of these criteria as possible. (If you cannot think of any more improvements, quit, and either reject or accept the final version.)
- Try out your modified version.
- Go back to step one, and repeat.

1.5 Desirable criteria for simple demos

Criteria for the design of good demos can be put under various headings, such as simplicity, pedagogical soundness, and good physics. (If I were a less modest person, I would suggest "Ehrlich's three laws" of demos: (1) Keep them simple, (2) Keep them pedagogically sound, and (3) Get the physics right.) Let's start with the criteria that fall under the simplicity heading.

Simplicity obviously means that the demo be kept as "low tech" as possible, consistent with getting the desired result. (But, of course, you do need to guard against using such low-tech apparatus that the observations become either ambiguous or too imprecise to be of any value.) Using items found around the house or office would be particularly desirable, because this makes reconstitution of the demo easy, if it should get lost, broken, or borrowed. Above all, simple demos should avoid expensive or hard-to-get parts.

Simple demos should make no use of such devices as oscilloscopes, air tracks, or computers—although, one of the demos in this book does suggest, half-jokingly, using a computer in a demo by dropping it! Meters, bulbs, and inexpensive easily purchased electronic parts, as well as parts found around most physics classrooms are all acceptable, with the laser being just borderline acceptable—partly because of the increased availability of compact laser pointers.

Compactness and portability are highly desirable features in a simple demo, as is the ability to set up, perform, and take down the demo in very little time, and to store it conveniently after use. Safety is an important consideration—despite the obvious appeal of demos that appear to put the demonstrator's life in danger. Safety might not seem to belong under the general heading of simplicity, but then again lawsuits could make your life anything but simple, if someone should get hurt emulating your demo.

The demo should be highly visible (or audible), and leave little doubt as to the actual outcome. Demos that use the OHP are particularly useful, but it would be best if they didn't *require* the OHP, so that they could be redone at home. Most important, a good demo should work every time, assuming you do it right. The very best demos, however, work every time, even if you don't "do them right"—that is, they require little by way of luck, manual dexterity, carefully crafted apparatus, or special environmental conditions.

Simple demos need not be simple in terms of the underlying physics, but they should have some kind of plausible qualitative explanation without your resorting to advanced physics. Good simple demos may be highly quantitative, but they should involve the smallest number of measurements to achieve the desired result. The simplest types of measurements are also the best, and best of all would be the null result—particularly when a null result is counterintuitive. Finally, simplicity involves the hard to define, but easy to recognize aesthetic dimension.

1.6 Good pedagogy

Apart from simplicity, good demos should also be pedagogically sound—meaning that something meaningful should be learned from them. Of course, the pedagogical soundness of a demo depends at least as much on how it is used in a class setting as on the qualities of the demo itself. One physics professor of my acquaintance noted that many students in his classes seem to stop paying attention when the demos are wheeled into the classroom—apparently these students believed the demos to be simply an entertainment break from the material for which they will be held responsible on their exams. The attitudes of this professor's students toward demos seemed to change markedly when they were better integrated into the lecture material, and when an occasional exam question related to a demo.

Moreover, this negative student reaction toward demos does *not* appear to be the typical one. According to a survey by R. Di Stefano, [3], most students view demonstrations very positively. In fact, student surveys on the value of demos show that demos help them to visualize and think about physics. (In contrast, student comments about the entertainment value of demos, and the way they break up the lecture were relatively infrequent according to Di Stefano's survey.)

In any case, here we are primarily interested in how the *intrinsic* qualities of demos can make positive or negative contributions to their pedagogical soundness. Foremost among the properties making for a positive contribution is simplicity. For example, a low-tech demo without hidden parts has an important pedagogical advantage over one using some "black box" apparatus that may seem remote from one's everyday experiences. Using demos that are literally transparent, and can be seen on the OHP, helps to emphasize the absence of hidden parts. Moreover, if students repeat your low-tech "home doable" demos on their own, they can have a lot of fun (and may even learn something), while showing them to their friends and relatives.

Good demos should serve as a vehicle for group discussion and debate. If the outcome of the demo is not obvious—or even better, counterintuitive—useful discussion could take place both before the demo as well as afterward. Discussions before the demo are facilitated by casting the possible outcomes of the demo in terms of a discrete set of possibilities. For example, you might ask: "if we quadruple the mass hanging on a slinky, will the period of oscillations be quadruple, double, half, or the same as the original period?"

Such discussions can, of course, also occur if you cast the demo in its qualitative form. In case of the mass-spring example, before doing the demo, you might simply survey the audience as to whether they predict adding more mass on the slinky will increase or decrease the period of oscillations. Demos that leave room for variations and extensions—possibly spontaneously suggested by the audience during the demo—are particularly valuable.

For example, with the mass on a slinky demo, it becomes only natural to ask what the effect on the period would be when other variables are changed, such as the length of the slinky or the size of the oscillations. One advantage of such simple demos as a mass on a slinky is that instant replays are easy, given that the demo takes no set up time, and very little time to perform. Some demos even lend themselves to slow-motion instant replays.

1.7 Undesirable features

Many demos are much more dramatic than the humble mass on a slinky, and there is great value in showing memorable demos that will capture the imagination of the audience, and long be remembered. But, our hope is that the memory will be focussed more on the physical principles behind the dramatic demo, and not the nuttiness of the demonstrator to try such foolishness. If this hope is misplaced, then our use of dramatic "gee-whiz" demos in a class setting may serve to entertain, and even put a human face on the physics, but they may not result in much learning, and they may even reinforce misconceptions.

There is nothing wrong with demos that may risk making a fool out of the teacher, but you should at all costs avoid demos that have the potential of humiliating a student. One of the demos in *Turning the World Inside Out,* involved your throwing raw eggs at a sheet held by two people.[4] The eggs should not break no matter how fast you throw them at the sheet for the same reason that an air bag protects you in case of a car crash: by stretching out the time of collision, the size of the decelerating force is reduced. Once when doing the demo in class with students holding the sheet, I came quite close to missing the sheet and hitting a student. I have not done the demo using students holding the sheet again, nor would I be comfortable being the one to hold the sheet for a student egg thrower.[5]

It is also probably unwise to do demos that trick the audience. For example, in the well-known demo in which a can of water is whirled at the end of a string in a vertical circle, the water remains in the can as long as the can has a linear speed at the top of the circle that exceeds some minimum value ($v_{min} = \sqrt{gr}$). The idea is that for velocities less than $\sqrt{gr}$, the centripetal acceleration at the top $a = v^2/r$ will be less than g, and the water falling with acceleration g will therefore become separated from the can. I have sometimes seen this demo done using a sponge that soaks up all the water in the can, so that no water pours out even at the end of the demo when the can is turned upside down.

Such unnecessary tricks can only breed student suspicions about the reliability of the demonstrator, and they blur the line between physics demos and magic tricks. On the other hand, there are some rare situations when a "trick" can lead to interesting discussions of new physics. For example, consider the well-known demo where a card is placed over the opening of a water-filled glass or bottle which is then inverted. The water does not fall out because if the card flexes slightly due to the weight of the water, the resulting small difference in air pressure between the inside and outside air is enough to hold up many centimeters of water.

One startling variation of this demo involves using a glass in which a fine wire mesh (from a window screen) has been glued just inside the top opening of the glass. This mesh offers little resistance to water when it is poured into the glass, so most of the audience will not notice it. However, because of the mesh, you can actually remove the card after the water filled glass has been inverted, and still not have the water pour out. Surface tension keeps the water in the inverted glass—at least as long as the glass is not tipped away from the vertical orientation. If you have managed to keep the presence of the mesh a secret, this

How to Design Simple Physics Demos

demo never fails to elicit a gasp from the audience. In view of this shocking finale, and more important, the interesting physics you can discuss connected with the mesh-covered glass, I am willing to exempt this clever demo from my admonition against demos that trick the audience.

For some topics in physics—especially topics in modern physics—the possibility of doing real-life demonstrations involving simple apparatus is out of the question. Even if you have the use of more elaborate apparatus, doing a live demonstration in an area such as time dilation in special relativity is not feasible. In such cases, analogy or simulation demos are quite acceptable—see for example the simulation of the Michelson-Morley experiment (number 12.8), and the one on time dilation (number 12.6). In areas of physics where the "real thing" can be shown, however, you should *usually* avoid simulations.

There may be exceptional cases where a simulation or analogy demo is worth showing even when the real thing is available—particularly where the simulation provides insights not easily otherwise obtained. One example would be the simulation of interference between two point sources of waves using the overlaying of two sets of concentric circle transparencies on the OHP described in *Turning the World Inside Out.*[6] The so-called Moiré patterns produced in this demo are virtually identical to a snapshot in time of a two-source interference pattern. Moreover, the ease of varying the source distance, and observing instantaneously its effect on the directions of maxima and minima in the intensity pattern, make this simulation of interference worth doing, even if many real interference demos are also available.

Demos having no clear connection to everyday situations are also undesirable. Sometimes, it takes only a little imagination to model a real-life situation in a way that leads to an interesting demo. For example, consider the story involving a large boulder in a row boat. The question we wish to consider is whether the level of the lake rises, falls, or stays the same when the boulder is thrown overboard? This "real-life" situation can be modeled using a styrofoam cup (the boat) floating in a container partly filled with water (the lake), with some weights (the boulder) placed inside the cup so that it just barely floats. If the container diameter is not much larger than that of the cup, the drop in water level in the "lake" when the "boulder" is thrown overboard is quite noticeable. In fact, a "slow-motion" repeat of the demo should makes it very clear to observers why the outcome occurs. Obviously, this demo has greater appeal when put in the context of a real-life floating boat in a lake—although you should note that in the case of a real lake the drop in water level would be miniscule.

The most pedagogically unsound demos are those that reinforce student misconceptions. For example, one demo I have seen uses a strong permanent magnet mounted over the edge of a desk with a second small magnet tied to a long string anchored to the floor. The magnet on the string is positioned a small distance below the stationary magnet. Due to the mutual attraction of the magnets, the small one appears to float beneath the large one, while keeping the anchoring string taut.

When you bring steel scissors nearby and close the blades in the space between the two magnets, explaining that you are "cutting the magnetic lines of force," the magnet on the string is found to fall to the ground at the instant you close the scissor blades. Unless the demonstrator explicitly makes clear that the demo is really a joke, the "explanation" accompanying it would very likely reinforce an incorrect view of lines of force as being real physical entities, and it masks the real reason that the magnet falls. Hint: would you expect the magnet to fall if you used a pair of scissors not made of steel to "cut the lines of force?"

Below I have outlined a summary listing the previously discussed criteria for a simple pedagogically sound demo, with the "others?" at the end merely serving as a reminder to add your own criteria to these two lists. Furthermore, you may find it useful to identify in some way those items on each list that you find particularly important or not important at all.

Keep it Simple
- Inexpensive
- Low tech
- Avoids hard-to-get parts
- Uses items found around the house
- Little or no construction required
- Not fragile
- Easily "reconstitutable" if lost

- Compact and easily storable
- Portable (ideally, pocket sized)
- Minimal set-up and take-down time
- Short execution time
- No mess afterward
- Safe as possible
- Outcome highly visible/audible
- Can be done with or without OHP
- Works reliably (if you do it right)
- Works reliably (*even if you don't!*)
- Minimum number of measurements
- Simplest possible measurements
- No advanced math/physics needed
- Aesthetically pleasing

Keep It Pedagogically Sound
- Low tech (no hidden parts)
- Home doable
- Literally transparent (OHP use)
- Good discussion possible
- Discrete or quantitative outcomes
- Surprising or counterintuitive
- Memorable—but not "gee whiz"
- No potential student humiliation
- No deceptive "tricks" or gimmicks
- Open ended—many variations
- Qualitative and quantitative
- Instant replays easy
- Multiple senses engaged
- Model for a real-life situation
- Illustrates important topics
- Doesn't reinforce misconceptions!
- Avoids simulations
- Others?

1.8 Getting the physics right

In addition to being simple and pedagogically sound, it is equally important that when you show a demo that you understand the physics behind it, *and* that the demo illustrates the principle you think it illustrates. A favorite simple demo of mine, for example, is the "radiometer"—a glass globe with four vanes having black and mirrored sides that turn on a low-friction mounting when it is illuminated with a bright light. Why does the radiometer turn? Based on conservation of momentum, we know that a ball which bounces off a wall delivers a greater momentum to the wall than another ball of the same mass and velocity that sticks to the wall. In fact, for a very light elastic ball which bounces backward with its initial speed, the momentum delivered to the wall is twice as great as the ball's original momentum. In just the same way, we expect that a photon of light gives more momentum to a surface when it is reflected than when it is absorbed. Therefore, light should exert a greater pressure on the mirrored sides of the vanes than the absorbing black sides.

I had been using the radiometer to illustrate these ideas, until one semester an observant student noticed that the vanes were turning the wrong way, as though the blackened side of each vane was experiencing the greater light pressure instead of the mirrored side. I still use the demo when discussing the radiation pressure of light, but at least now I know enough to explain that the effect of light pressure is overwhelmed by another one having the opposite sign: the greater heating of the blackened sides.[7] This greater heating causes residual air molecules in the globe to recoil more forcefully when they hit the black sides, thereby giving them a greater impact than the mirrored sides.[8]

Instructor misunderstandings concerning the basic physics behind a demo are more common than you might think, and extend to some very popular demos. For example, a favorite demo of the concept of inertia involves quickly pulling a table cloth out from under a table setting, which, under optimum circumstances, remains relatively undisturbed. But, this demo really illustrates the equality of impulse ($F\Delta t$) and change of momentum ($m\Delta v$), rather than "inertia." The demo shows that a given force (the friction of the table cloth), has little effect on the velocity of the table setting, only insofar as the time it acts, Δt, is kept short. Clearly, the outcome of the demo would be quite different if the cloth were removed slowly, and the same force acted for a longer time.

So, be sure that you can explain the physics behind a demo, and that you can account for the result, even when the result is unexpected. Explaining an unexpected result as being due to some other well-understood physics is not as problematic as "explaining away" a result arising from unknown causes—which can only increase student's suspicions that "if it doesn't work, it's physics!" On the other hand, showing demonstrations that don't work properly the first time, but do work after some minor adjustments suggested by students, can be an especially effective classroom strategy. Such demos are much more likely to be remembered than uneventful successful ones.

1.9 A case study in improving a demo

The process of taking a good demo and making it better can be illustrated through a specific example. A well-known demo in the area of rotational dynamics is to roll solid and hollow spheres and cylinders down an incline, and have the audience predict which objects should reach the bottom of the incline first. In such "races," the objects having the smallest moment of inertia (for a given mass and radius), should have the highest acceleration of their center of mass. Therefore, the predicted order of finishing would be: solid sphere, solid cylinder, hollow sphere, and hollow cylinder, independent of their mass or radius.

This demo already has many desirable features of a good simple demo. It is low tech, safe, and inexpensive, and requires no construction or extensive set up or execution time. The demo also usually works, is unobvious, and allows for student discussion before and or after the demo. It also has various possible extensions, such as rolling cans with various amounts of liquid down an incline.[9] One drawback in the way the demo is usually performed is that it is not quantitative, although it might only be a drawback if you wanted to use the demo at the high school or college level. In any case, this drawback could, in theory, easily be remedied, given our ability to predict velocities for each rolling object at the bottom of the incline. For example, if we use for the moment of inertia of a rolling object $I = kmR^2$, it can be shown, using conservation of energy, that the velocity at the bottom of the incline in the absence of rolling friction is given by

$$v = \sqrt{\frac{2gy}{1+k}}$$

where the k values are 2/5 for a solid sphere, 1/2 for a solid cylinder, 2/3 for a hollow sphere, and 1 for a hollow cylinder—assuming that the hollow objects have very thin shells, and the solid objects have a uniform density. If we let v_s be the velocity of the solid sphere, the velocities of the other shapes are therefore predicted to be $0.837v_s$ for a hollow cylinder, $0.966v_s$ for a solid cylinder, and $0.916v_s$ for a hollow sphere.

You might imagine testing these predictions by timing individual rolls down an incline of specific length, and then comparing the objects' relative speeds. But, the comparisons can be accomplished with far fewer and simpler measurements. For example, if a solid sphere rolls a distance L in a certain time, then ignoring rolling friction, a hollow cylinder, solid cylinder, and hollow sphere should roll distances $0.837L_0$, $0.966L_0$, and $0.916L_0$, respectively, in the same time, based on the previously quoted velocities.

This type of comparison—essentially a null result—can be made by "eyeball," especially if the incline slope is not too large, and the motions are relatively slow. The test of the null result can also be made audible as well if the two rolling objects hit a "stopper" board at the bottom. (Such a refinement might better serve to check on whether the finishes are simultaneous, because the human ear can easily tell if a time interval exists between two sounds, and in principle, you should engage as many senses as possible in a demo.) If you find that a pair of objects do not finish their race simultaneously when one is given the lead that theory predicts, try varying the lead until the finishes are simultaneous.

But, a most important part of improving a demo is to try out the improvement. If you actually try a random pair of rolling objects, you are unlikely to get anything like the results just mentioned! In fact, you are likely to find very significant differences in rolling distances for a simultaneous finish when using pairs of objects having the *same* shape—differences that depend on the mass, radius, and material composition of the rolling object. Presumably, these differences can be traced to the effects of rolling friction, and slight nonuniformities in shape, which become most significant for small m, R, and θ, and especially for cylinders rather than for spheres.

After some experimentation, you should be able to find some examples of each shape which exhibit minimal rolling friction—the fastest finishers of each shape. For example, using solid and hollow cylinders having the smallest rolling friction I could find, I obtained a measured distance ratio, 0.875 ± 0.013, for simultaneous finishes, which is quite consistent with the theoretical value, 0.866. This ratio corresponded to giving the hollow cylinder a lead of $0.23 \pm .02$ meters when the solid one rolled a distance of 2 meters. (Measured comparisons between a solid sphere and solid cylinder were not found to be particularly close to theory, presumably because of the greater frictional effects for cylinders compared to spheres.)

Further improvements in the demo relate to its portability and visibility. For example, you normally would want to use a long inclined board to run the race, in order to give long races, and easily distinguishable outcomes, but that choice would present problems in portability, unless you had a readily available table which could be propped up at one end. If you wanted only to illustrate the qualitative feature of the demo—showing which of two shapes finishes the race first—you could dispense with the board entirely, and do the demo on an overhead projector, one side of which has been propped up to create an incline. A relatively small angle of incline will keep the rolling times long, and easy to compare, but if the angle is too small, the cylindrical shapes will not roll if they have any slight nonuniformities.

1.10 Importance of being quantitative

Our suggested improvements in this demo have focussed primarily on making it more quantitative. In fact, with many of the demos throughout the book we have been as quantitative as possible—perhaps more quantitative than many readers would care to be. This emphasis on the quantitative aspects has several motivations. First, because there is a certain element of delight in seeing how very simple equipment can give quantitatively correct results. But, even when the results are not extremely precise, it is worth remembering that physics is a quantitative science, capable of making specific numerical predictions that can be tested.

Quantitative versions of demos may be valuable even when you are not testing a specific prediction. Perhaps instead you can use it to *formulate* a relationship when none is known—a much more creative exercise. The ideas of scaling and power laws can sometimes allow us to formulate and test such relationships—see demos 2.13 and 8.8, for example. In fact, many of the demos in their quantitative form can be used either as mini-experiments in a lab setting, or as demos to test a specific calculation in a lecture or interactive seminar.

But, there is an even more important reason to be quantitative about a demo, namely, in its qualitative version the demo may not prove what you think it does. For example, in the normal qualitative version of the demo where various shaped objects are rolled down an incline, the result found is that spheres usually beat cylinders, and solid cylinders usually beat hollow ones. An issue left unresolved by the qualitative demo would be whether this result is due to differences in the amount of rolling friction for the specific objects used, or differences in their moments of inertia. In view of the large differences found for objects of a given shape—presumably due to rolling friction—*only* a check testing a specific numerical prediction using low friction objects can give us confidence that we are seeing results that reflect differences due to the object's moments of inertia.

In discussing ways to improve demos throughout this chapter, I may have conveyed the impression that it is always a good idea to improve a demo as much as possible before showing it to an audience. But, there may be occasions when it is useful to deliberately rig a demo so that it doesn't work so well, and to involve the group in a creative discussion of how it could be improved. Such a discussion would be particularly fruitful if the modifications could be carried out on the spot.

Notes

1. R. Ehrlich, *Turning the World Inside Out and 174 Other Simple Physics Demonstrations,* Princeton, NJ: Princeton University Press (1990).

2. The PIRA website is maintained by the University of Washington physics education group. See: www-hpcc.astro.edu/scied/physics/physdemo.html.

3. R. Di Stefano, The American Journal of Physics, **64**, 58–68 (1996).

4. R. Ehrlich, *Turning the World Inside Out*, p. 32.

5. Some instructors might use students both as egg throwers and sheet holders, but that might also prove highly problematic in the event of an "accident." Student conflicts have been known to arise starting from far less provocative situations. In addition, the instructor might end up with egg on his or her face!

6. R. Ehrlich, *Turning the World Inside Out*, p. 186.

7. The radiation pressure of the photons would become the dominant effect only if the pressure inside the radiometer bulb were less than a millionth of an atmosphere, according to an informal communication from Harry Morel of Arizona State University.

8. One nice way to demonstrate that it is the heating effect of the light that causes the vanes to turn is to illuminate the radiometer with a focused flashlight beam that passes through a layer of about 5 centimeters of water (in a transparent spaghetti holder, for example), before reaching the device. The radiometer vanes will not turn in this case, because the water filters out the infrared radiation that causes most of the heating. However, the vanes resume turning as soon as the water is removed. Another way to show that the vanes turn due to thermal effects was suggested to me by Ron Edge. If you put the radiometer in a refrigerator, when you take it out it spins on its own, because the blackened side of the vanes cools off faster than the mirrored sides.

9. Experiments rolling cans partially filled with liquid yield some surprising results, which may not yet be fully explained. See, for example, K. A. Jackson, J. E. Finck, C. R. Bednarski, and L. R. Clifford, The American Journal of Physics, **64**, 277–82 (1996).

Chapter 2

Newton's Laws

The demos in this book are grouped according to various important physical topics, such as Newton's laws of motion. Although the demos in this chapter all deal with Newton's laws, some could have just as well been included under other headings, such as fluids, or conservation of momentum and energy. The relative numbers of demos listed under various topics does not reflect any assessment of the topic's importance, rather it reflects my ability to find a set of demos for that topic that meets most of my criteria for a good simple demo discussed in the previous chapter, without overlapping the collection in *Turning the World Inside Out.*

In this chapter, no attempt has been made to separate demos according to Newton's first, second, or third laws, because in many cases a demo illustrates all three laws. The order of demos within each chapter is quite arbitrary, but easy ones are placed near the beginning, and those requiring more mathematical sophistication are placed toward the end.

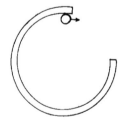

2.1 The missing circular arc

Demonstration
A ball rolling within a constraining circle flies off on a tangent if a section of the circle is missing.

Equipment
A marble or steel ball and a plastic ring of the type used in computer magnetic tapes, which can be obtained from most computer centers. An alternative would be an embroidery hoop.

Discussion
According to Newton's first law, an object on which no forces act travels at constant speed in a straight line, regardless of its prior motion. The present demo

illustrates this simple idea, and it explicitly confronts a common misconception regarding circular motion.

Place the plastic ring on a flat horizontal surface, such as the overhead projector (OHP), and roll the ball around the interior of the ring. Now, replace the ring with one that has a 90 degree arc cut out of it, and quiz the audience as to what will happen when the ball is rolled around the inside of the partial ring. You may wish to give people a choice of three specific possibilities, namely, that when the ball reaches the missing segment, it will: (*a*) fly off on a tangent line, (*b*) complete the circular orbit, or (*c*) fly radially outward. So as to avoid any confusion over terminology, you might show these three possible paths on a transparency placed under the ring. You may find that some people believe that (*b*) is the actual motion, perhaps based on their belief that the ball had a "memory" of its prior circular motion just before the missing arc.

A related demo would be to observe the path of a rock tied at the end of a string, when it is whirled in a circle fast enough for the string to break. However, for safety reasons, you probably would want to do a modified version of that demo, in which a small ball is very lightly taped to another object being whirled in a horizontal circle. Whirl the object and attached ball at increasing speed, and observe that when the ball flies off the object, it does so along a tangent to the circle. In my experience, many people who correctly predicted that the marble would fly off on a tangent in the partial ring demo, nevertheless seem to believe that the ball, when it breaks free, should fly radially outward, which of course is only true in a rotating frame of reference. (For example, an athelete in the hammer throw would see the hammer move radially outward after it left his or her hands.)

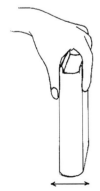

2.2 Estimating the net force on a moving book

Demonstration

Observers can try to estimate the net force acting on a physics book as you or they move it in several types of motions.

Newton's Laws

Equipment

A physics book.

Discussion

We can learn some physics from a physics book even without opening it—although, of course, the demo works using any object of comparable mass. According to Newton's second law, the net force on an object equals the product of its mass and acceleration, both of which can be roughly estimated by observers watching you move an object. Hold the book oriented vertically in one hand, and move it horizontally at constant velocity. Afterward, move it repeatedly back and forth horizontally through a distance of about a foot, completing each cycle in a time of about one second. (You may wish to use a metronome to count off the seconds.) In each case, challenge the audience to estimate the net force on the book based on their observations.

For the uniform velocity motion of the book, the net *horizontal* force of your hand is of course zero. For the back and forth motion, the net horizontal force is given from Newton's second law, $F = ma$, where m is the estimated mass of the book, which you may wish to supply. To get a numerical estimate of the acceleration, observers could assume its acceleration and deceleration is uniform during the back and forth motion. Specifically, it might be reasonable to suppose that during each half oscillation, the book is first accelerated a distance $s \approx 0.5$ feet (0.15 meters), during a time $t \approx 0.25$, second, and then decelerated over a similar distance and time. Observers should be able to estimate the acceleration and deceleration using $a = 2s/t^2$, which here gives $a \approx 16 \, \text{ft}/s^2 = 4.9 m/s^2$, or half the acceleration due to gravity.

An alternative demo would be for you to specify the desired motion of the book—either in words or by providing a velocity-time or acceleration-time graph—and challenge members of the audience to move their own book accordingly. One advantage of this way of doing the demo is that by repeating it using either larger distances, x, or smaller times t (and hence higher accelerations and decelerations), the need for larger forces to maintain the motion of the book becomes immediately obvious to the mover—which brings home Newton's second law in a most direct way *kinesthetically*. Newton's second law could be illustrated further by varying the mass, while keeping the motion the same as previously.

One virtue of varying the mass rather than the acceleration, is that you could then also use the demo to illustrate Newton's *third* law. For example, suppose a very light mass is held between your two hands and accelerated. The net force required to achieve the acceleration is negligible, according to Newton's second law. In other words, your right and left hands must exert nearly equal and opposite forces on a very light mass.

In the limit as the mass between your hands goes to zero, the net force on that zero mass object would be zero, meaning that your hands exert equal and opposite forces on *one another,* even though one hand is pushing the other aside. This difficult-to-grasp application of Newton's third law may be easier to accept when you go through the process of using smaller and smaller masses between your hands, than if you had simply started with one hand forcefully pushing the other aside, and asked someone to compare the forces each hand exerts on the other. (In that case, many people would probably mistakenly say that the hand that pushed the other one aside exerted the greater force.)

2.3 Force, mass, and acceleration

Demonstration
Using a very simple force indicator attached to your two fingers, you can move an object in various ways, and show how the net force on it depends on both its mass and the manner it is moved.

Equipment
Pieces of folded index cards, a transparent plastic grooved ruler, a one-inch (2.54 cm)-diameter ball made of stainless steel, and a second one made of styrofoam. You may also wish to prepare a hollowed-out styrofoam ball with a steel center. Styrofoam balls can be obtained at crafts stores, and a small steel ball can easily be pressed into its center if the styrofoam ball is cut in half.

Discussion

In the last demo the audience could kinesthetically feel how the net force on an object depended on its motion. In the present demo, using a primitive "force indicator," they can observe the connection between force and acceleration *visually*. All you need to make the force indicator is a piece of index card. Simply cut two pieces from the card, having approximate dimensions one by four centimeters. Fold each piece in half making a V shape, and tape one side of each V shaped piece to your two index fingers, or between your thumb and index finger of one hand. The net result is that if you close your fingers with the V's between them, the opening angle of each V decreases as you press them together. Now, if you place an object between your fingers on the OHP, the easily seen opening angle of each V resting against the two sides of the object is an approximate measure of how hard each finger presses on it.

Place a one-inch (2.54 cm)-diameter steel ball in the groove of a plastic ruler on the OHP. Move the ball back and forth in the groove while it is between the index cards taped to your fingers. Observe how the opening angles of the index card V's change during the periods of acceleration and deceleration of the ball, as the ball is moved from one point on the ruler to another. You should find that, as a result of a sudden acceleration or deceleration, the ball oscillates back and forth between your fingers, since the two folded index cards behave much like gentle springs. But such oscillations of the ball occur only during periods of acceleration and deceleration, such as those at the two ends of its motion along the ruler. During the constant velocity portion of the ball's motion, it should remain roughly equidistant between your fingers, meaning the opening angle of the two V's should be equal—which is the case in the figure.

Now, replace the steel ball by a styrofoam ball, and repeat the experiment. In view of the small mass of the styrofoam ball, a very small net force should be needed to accelerate it. Therefore, the ball should stay equidistant between your fingers, regardless of whether or not it moves with constant velocity. In other words, for a very small mass, the net force on the ball—as measured by the difference between the opening angles of the two index card V's—should remain quite small even if the ball is accelerated. This demo shows the dependence of net force on the mass being accelerated.

A possibly more interesting way to do the demonstration would be to use two balls that outwardly look the same, but have very different masses, for example, one made of styrofoam and another styrofoam ball having a steel ball embedded inside it. In this case, when you repeat the demonstration using the two identical looking balls, observers watching the behavior of the two index cards as each ball is moved should easily be able to conclude which ball has the larger mass.

2.4 Picking yourself up by your bootstraps

Demonstration
Try to pick yourself up by your bootstraps or belt.

Equipment
None.

Discussion
The English expression inviting people to "pick themselves up by their bootstraps" is an exhortation to them to take the initiative. But, of course, literal attempts to pick yourself up by your bootstraps (or belt) would seem to be doomed to failure, because only an *outside* force can give you an upward acceleration, and cause you to leave the ground. You can explain all this to your audience, while frantically tugging on your belt. It might be amusing, if after a while, you jump off the ground, timing your jump with an especially hard tug. (Do it while flexing your knees only slightly, so as to make the jump a surprise to the audience, maybe even pretending to be shocked yourself.)

Unfortunately, jokes can be misunderstood! So if you are worried about the possibility that the audience may be left with the impression that a large enough internal force can, in fact, lift you off the ground, you may wish to explain the true cause of your leaving the ground as being due to your "yoga flying" ability. If that explanation is taken seriously, you had better quickly explain that the real reason for your sudden liftoff was that by rapidly changing your mass distribution, you caused your feet to press down with a force greater than your weight, and the reaction force of the ground pushed back on your feet with a force greater than your weight.

Unfortunately, this explanation itself can lead to misunderstandings as well, since you need not leave the ground just because the upward force on your feet *momentarily* exceeds your weight—that would only be the case for a rigid body, which you presumably are not. You would only lift off the ground if the upward force of the ground was large enough, and lasted a long enough time. The closer you come to being a rigid body, however, the easier it is for you to achieve "liftoff" during a short time interval, because your center of mass needs to accelerate only a short distance upward to leave the ground.

Surprisingly, for you to leave the ground, it is neither necessary nor sufficient that the upward force of the ground on your feet exceed your weight. As shown in the next demonstration, the force that your feet exert on the ground (or a scale) can exceed your weight even though they don't leave the ground, anytime your center of mass accelerates upward, or decelerates downward. In addition, it is actually possible to leave the ground without the ground exerting a force on your feet that exceeds your weight. In order to accomplish this feat, simply lift your feet off the ground by bending your legs at the knees and quickly pulling your lower legs up, rather than pushing off against the ground.

In such a case, your feet leave the ground, not because of the force of the ground on your feet, but because your feet were pulled upward by the rest of your body before it had a chance to fall appreciably. If you try this maneuver while standing on a scale, the scale reading should not exceed your weight just before liftoff, and it could even be less than your weight, depending on the motion of your center of mass. You probably would want to straddle the scale as your feet land, so as not to damage the scale.

2.5 Deep knee bends on a bathroom scale

Demonstration
You can make a rough test of Newton's second law by doing deep knee bends on a bathroom scale.

Equipment
A bathroom scale.

Discussion
If you do deep knee bends on a bathroom scale, the scale reading will vary because your center of mass (CM) accelerates during your motion. As you begin the deep knee bend, your CM accelerates downward for a while, after which it moves with roughly constant velocity, and finally it accelerates upward, as you decelerate to rest at the lowest point. It might be instructive to have someone watching you do the deep knee bend estimate the approximate variation in the scale reading, based on their estimate of the amount your center of mass descends, and how long they perceive the periods of acceleration and deceleration to last. You could then announce the experimental results after the predictions are made.

Using Newton's second law, it can easily be shown that during the initial phase, when the downward acceleration is a_i, the scale should read *less* than your weight by the fraction a_i/g, and during the final phase, when the deceleration is a_f, the scale should read *more* than your weight by the fraction a_f/g. For example, assuming that $a_i = a_f = g/10$, then if you weigh 150 pounds, the predicted scale readings should range from 135 pounds as you begin the deep knee bend to 165 pounds toward the end. Furthermore, if we assume the acceleration of your CM is uniform as it moves downward a distance y in a time t, we can find its value using $a = 2y/t^2$.

Thus, for example, suppose that during a deep knee bend your CM moves downward by about half a meter during the deceleration stage, then that stage would have to last about a second to achieve a final acceleration of $g/10$. You may want to focus primarily on observing your weight during the deceleration, because with a scale whose dial is rapidly changing, it turns out to be somewhat easier to observe the final scale reading than the initial one, since in practice, the initial acceleration is usually much greater than the final. (The acceleration is probably going to be larger than the deceleration, because you are just relaxing your muscles and letting gravity do the work to begin the descent.)

2.6 A "Monkey-Hunter" variation

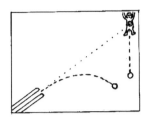

Demonstration
The classic "Monkey-Hunter" demo can be done using only a transparency on the OHP.

Equipment
Two one-inch (2.54 cm)-diameter stainless steel balls, a transparency blank, and several popsicle sticks.

Discussion
In *Turning the World Inside Out,*[1] I suggested a version of the classic "Monkey-Hunter" demo suitable for the OHP. However, that demo did require a bit of construction and some special materials (a lucite sheet). The version described here can be made in under five minutes, and requires no special materials.

Turn the transparency blank so the long dimension is horizontal, and draw a picture of a monkey hanging from a tree in the upper right corner of the transparency. (Actually, in today's politically correct times, you may want to have the monkey shooting at a hunter hanging from the tree!) Tape two pieces of popsicle stick (about 5 cm long) onto the lower left part of the transparency with a 0.5 cm spacing between them all along their length forming the "gun barrel." The sticks should be oriented so that the gun barrel points directly at a bull's-eye on the picture of the monkey (or the hunter, if you prefer).

Tape the completed transparency onto the OHP, and prop the back end of the projector up by a few centimeters. Verify that the orientation of the transparency is correct by allowing a ball to roll from rest starting at the bull's-eye, and checking that it rolls parallel to a the right edge of the transparency. Now, place one ball in the gun barrel and the second ball on the bull's-eye on the monkey. Simultaneously propel one ball along the barrel and release the other one on the bull's-eye, and you should find that they collide after the former ball has traveled some distance along a parabolic arc.

Achieving the desired result does take a bit of hand-eye coordination, but with some practice I can usually make the balls collide about two-thirds of the time. The main trick is to release the balls simultaneously, and to hold onto the ball leaving the gun until it is near the end of the barrel. You may also want to try varying the tilt of the projector.

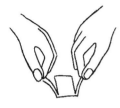

2.7 Tearing a card into three pieces

Demonstration
If you challenge someone to pull on the two ends of a piece of paper that has been nearly cut into three pieces, they will probably not be able to pull them off simultaneously, so as to leave a severed middle piece.

Equipment
A piece of paper or an index card which has two cuts that almost cut it into three pieces.

Discussion
The inability of someone to pull two nearly severed pieces of paper off at the same time is quite surprising to most people. To set up the demo, cut the two end pieces so as to leave no more than a millimeter thickness of paper connecting on each side to the middle piece. Even though the two cuts in the paper or card are made the same length, one side will invariably be slightly easier to tear than the other. Therefore, equal forces applied to the two sides will lead the easier side to break first. To avoid this outcome, someone might futilely try to pull the two side pieces so quickly (with high enough acceleration), so that the middle piece is not able to keep up with either one. But Newton's second law makes that outcome impossible to achieve in practice, because the mass of the card is very small, so that very little force is needed to accelerate it in one direction or the other.

On the other hand, there are a couple of "tricks" to achieve the seemingly impossible—one being more of a cheat than the other. The bigger cheat is to bite your teeth on the middle piece holding it in place while pulling the two end pieces. The slightly more respectable method of tearing the paper into three pieces simultaneously—since it relies on a physics principle—is to add enough mass to the middle piece of paper, perhaps in the form of pennies taped on the card. When you quickly pull the side pieces apart with enough added mass on the middle piece, the middle piece is now unable to follow either side piece, and it tears off immediately. This outcome will be more likely to occur if you pull the side pieces quickly than slowly.

We can explain the outcome in Newtonian terms by noting that the added mass on the middle piece allows us to pull the two side pieces with enough acceleration, so that the force needed to accelerate the middle piece in either direction now exceeds that needed to tear it loose. (In principle, it should be possible to do without the added mass to the middle piece if the two side pieces were each given a large enough acceleration, which apparently is not possible using simple hand motions.)

2.8 Timing the fall of dropped objects

Demonstration
By dropping objects and timing their fall through a known distance with a stopwatch, you can obtain surprisingly accurate measurements of the acceleration of gravity.

Equipment
A stopwatch accurate to one-hundredth of a second.

Discussion
It is commonly believed that, in view of a typical person's reaction time of 0.2 seconds, measurements made to time the fall of objects with a stopwatch are capable only of providing crude values of the acceleration of gravity. However, fairly precise values of g (accurate to better than a few percent), can in fact be obtained, if you take the proper precautions, and average the results of many measurements. For a full discussion of these precautions, see my article in the Physics Teacher.[2]

The basic experiment involves timing the fall of a coin (I used a nickel), from heights of 0.6, 0.9, 1.2, 1.5, and 1.8 meters marked on the wall with tape. Even though the experiment is of the low-tech variety, accurate results are highly dependent on using proper experimental technique. In particular, it is important to position the coin with its bottom edge even with the tape, and view it from the same height it is dropped to avoid parallax. Such precautions are needed, because for the lowest height, an error in initial height of only 0.6 cm would cause a 1 percent error in the value of g. (For drops at the lowest height of 0.6 meters it may seem that you are stopping the watch immediately upon starting it, and that you are not really measuring anything—although you are.) If working with a partner, you can obtain more accurate timings if the same person who releases the coin also times its fall—that way, you can anticipate both the moment of the coin's release and impact.

Suppose you were to measure repeatedly the time of fall t for drops from a given height, obtaining values $t_1, t_2, t_3, \cdots, t_N$, with a mean value $\bar{t}$. One way to judge the random component of the uncertainty in a measurement, Δt, is the standard deviation about the mean. If the measurements have a normal (Gaussian) distribution about the mean, they should lie within one standard deviation of the mean about 68 percent of the time. The standard deviation may be approximated from the measurements themselves using

$$\Delta t = \sqrt{\frac{\sum_{i=1}^{N}(t - \bar{t})^2}{N}}.$$

My standard deviation in times for 200 drops of a coin from a height of 1.5 meters was 0.037 seconds— far less than my reaction time. However, the "bad news" is that the average recorded time for the 200 drops was systematically off by -0.034 seconds from the value predicted assuming g has its standard value. The negative sign simply means that on the average I tended to stop the clock 0.034 seconds too early. But, suppose that for every person who tends to measure time intervals too short, like me, there is one who tends to measure them longer than their true value by the same amount. In that case, we could greatly reduce this source of systematic error by averaging time interval measurements for many observers.

One semester, I had the students in three separate classes each measure the fall of a coin. Each student was asked to record ten drops from the five heights mentioned above, rejecting only obviously incorrect times. By "obviously incorrect," I included cases where the measured times either were many standard deviations from their average value, or decreased with increasing height. The data for all students in a class were combined for each height to get an average time of fall for that height, and these averages were compared to the prediction $t = \sqrt{2y/g}$, with g having its standard value.

In over half the cases, the class averages for each height gave values within 3 percent of the value predicted—confirming the constancy of the acceleration of gravity, as well as its standard value. Clearly, air resistance appears to have had little consequence for drops from a height of 1.5 meters or less of a coin as heavy as a nickel. The average value of g for the three classes for all heights was within 1 percent of the predicted value! The computation of this average g involved a total of 6,000 separate measurements (3 classes, 40 students per class, 5 heights per student, 10 measurements at each height per student). The closeness to the predicted value of g supports the assumption that averaging times over many observers largely eliminates systematic errors due to some people tending to measure time intervals as being either too short or too long.

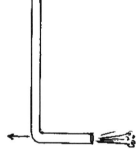

2.9 Recoil force on a bent straw

Demonstration
When you blow hard through a bent straw, the straw recoils just like a lawn sprinkler, but no recoil is observed if the straw is placed in a small plastic sandwich bag, or "baggie."

Equipment
A flexible soda straw with a 90-degree bend.

Discussion
There are lots of demos for Newton's third law of action and reaction, perhaps none simpler than that using a flexible soda straw. As noted, for example, in *Turning the World Inside Out*,[3] when you blow hard into the long part of the straw suspended vertically, the free bottom end recoils—thereby illustrating Newton's third law. The idea is that the air leaving the straw must experience a sideways force in order for it to make the right-angle turn, and the reaction to that force pushes the straw in the opposite direction. (If anyone should accuse you of using your tongue

to make the straw move, just invite them to try the trick.)

A nice twist to this simple demonstration was suggested by Professor Richard Solomon of the California Institute of Technology. The idea is to attach a small plastic bag at the end of the straw with a "twister tie," so that the air you blow through the straw is then confined to the bag. The air still exerts a sideways force on the straw as it makes its right-angle turn, but now it also exerts a canceling force on the attached plastic bag when it leaves the straw—so there is no recoil. Alternatively we can explain the lack of recoil as being due to the absence of any outside force on the straw, since the air doesn't leave the "system" of the straw plus baggie.

2.10 Magnet symmetry and Newton's third law

Demonstration
Two small equal mass horseshoe magnets held together with like poles touching, will fly apart when released into a symmetric configuration, confirming Newton's third law.

Equipment
Two small horseshoe magnets, which can be purchased for under ten dollars from any scientific equipment company, or hardware, nature, office supply, or educational supply stores.

Discussion
The use of two small horseshoe magnets to illustrate Newton's third law was described by Robert Chasnov and Louis Overcast.[4] As noted in that article, the demonstration can be given on an overhead projector if you use a transparency made from a sheet of graph paper with ruled coordinate axes, and position the two magnets at the origin of the coordinate system prior to their release. Upon their release, the magnets travel a short distance due to their mutual repulsion, and they always land in a symmetric pattern that usually shows evidence both of a mutual force, as well as a mutual torque. In other words, the magnets must fly apart with equal and opposite rotational

and linear velocities in order to land in a symmetric pattern.

You might want to try two variations on the demonstration. The first variation would be to add some mass to one of the magnets—perhaps in the form of a lump of clay. The added mass will of course alter the symmetric patterns, because equal magnitude reaction forces on both magnets no longer produce equal accelerations. A second variation might be of value in dispelling the common misperception that the stronger of two parties exerts a greater force on the other party than it feels back in return—which would be a clear violation of Newton's third law.

To make the previous point you could use two magnets, one of which has been deliberately weakened—either by heating it or hitting it with a hammer. In this case, the magnets will not fly apart as far as originally, but the patterns will still be symmetric, because weakening one of the magnets does not change the equality of action and reaction force magnitudes. (If you try this part—and it does seem a shame to ruin a good magnet—be sure not to reduce the magnetism of one magnet entirely, because the magnets will not exhibit any repulsion in this case, and in fact, will attract one another.)

2.11 Weighing a swinging pendulum

Demonstration
The apparent weight of a pendulum suspended from a spring scale varies during its swing in a predictable way, with the maximum apparent weight found at the bottom of the swing, and the minimum apparent weight at the two extremes of the swing.

Equipment
A large demonstration scale from which a one-kilogram mass is suspended by a string about one meter in length.

Discussion
If you have a rope that can just hold your weight, you would be in trouble if you wanted to use it to swing through the tree tops, because the tension will exceed your weight during the swing. This idea can be illustrated using a pendulum hung on a spring scale.

As you swing the pendulum, you will need to hold the scale only by a hook at its top, so as to allow the scale to swing freely along with the suspended mass—otherwise the mass cannot pull the spring in the scale properly. You will find that the amount of variation in scale reading depends, of course, on the range of angle through which it swings—with the largest scale reading found at the bottom of the swing, and the smallest at the two extremes.

We can easily predict the scale reading, S, at the bottom of the swing ($\theta = 0$), given that the range of swing angles is between $\pm\theta_0$. If the velocity of the pendulum of length ℓ is zero when its angle is $\pm\theta_0$, then by conservation of energy, we have $\frac{1}{2}mv^2 = mgy = mg\ell(1 - \cos\theta)$, so that the velocity at the bottom of the swing is given by $v = \sqrt{2g\ell(1 - \cos\theta_0)}$. As a result, the upward centripetal acceleration at the bottom of the swing is given by $a = v^2/\ell = 2g(1 - \cos\theta_0)$. Applying Newton's second law, we may write $\Sigma F = S - mg = ma = 2mg(1 - \cos\theta_0)$. Thus, the scale reading, or "apparent weight," is given by $S = mg(3 - 2\cos\theta_0)$. Equivalently, we can say that at the bottom of its swing, the pendulum's apparent weight is increased from its nominal value mg by the factor $3 - 2\cos\theta_0$. For example, if you hung a 1 kg mass, whose weight is 9.8 N, and allowed it to swing from an initial angle of 45 degrees, its apparent weight at the bottom of the swing would momentarily be increased by a factor of $3 - 2\cos 45° = 1.59$, for a predicted maximum scale reading of 15.6 Newtons.

The scale reading at the two extremes of the swing can also be easily predicted. At these two extreme points the pendulum is momentarily at rest, but it is accelerating in a direction perpendicular to the string. Thus, since there is no acceleration parallel to the string, the component of gravity along the string direction, $mg\cos\theta_0$, is just canceled by the tension in the string, T. The tension in the string is just what the scale reads, so we predict the weight mg to show a decreased scale reading $mg\cos\theta_0$ at the two extremes—a reduction from the nominal value mg by the factor $\cos\theta_0$, which would be a 13 percent decrease for a pendulum swinging between ± 30 degrees, for example. Even though none of these results depend on the length of the pendulum, a long pendulum is desirable, so as to slow down the variations in scale readings during the oscillation. In order to keep the swings of the pendulum in a well-defined angular range (say ± 30 degrees), you may wish to draw lines on a blackboard directly behind the swinging pendulum.

2.12 Weighing an hourglass

Demonstration

Contrary to popular belief, the weight of an hourglass actually *does* change as the sand flows, as can be verified in an equivalent demonstration in which the weight of water flowing from one bottle to another yields an increased scale reading during the time that the water flows.

Equipment

Two small plastic soda bottles, and a digital scale accurate to 0.1 grams, which can be found in most introductory chemistry laboratories. You may also want to use a "tornado tube" which may be obtained from some science or nature stores.

Discussion

It is commonly believed by many physicists that the weight of an hourglass does not change when the sand flows—see, for example, Jearl Walker's *The Flying Circus of Physics*.[5] This belief rests on a simple calculation, which shows that the extra force of impact of the descending sand is just compensated for by the weightlessness of the sand during its period of fall. The calculation assumes that some sand of mass dm falls from rest, and acquires a velocity $\Delta v = \sqrt{2gy}$ after falling for a time $\Delta t = \sqrt{2y/g}$. The time-averaged force that the falling sand exerts during its period of fall and subsequent impact can then be found from $F = m\frac{\Delta v}{\Delta t} = m\frac{\sqrt{2gy}}{\sqrt{2y/g}} = mg$, which is precisely the same force as if the sand had been at rest in the hourglass. However, despite this "derivation," the belief that the weight of the hourglass is unaffected by the falling sand is, in fact, mistaken!

The flaw in the preceding derivation is the incorrect assumption that the falling sand begins its free fall descent with zero velocity. As K. Y. Shen and Bruce Scott have shown[6], the center of mass of the falling sand moves downward, but with decreasing momentum, and therefore it actually accelerates *upward* as it moves downward. A proper derivation shows that the scale reading during sand flow would be expected to increase by an amount $\Delta F = v\frac{dm}{dt}$, where v is the velocity of the sand surface at the top of the hourglass and $\frac{dm}{dt}$ is the rate of mass flow.

The basis of this formula can be understood by considering the change in the distribution of the sand in a short time dt. In this time a small amount of sand dm having an initial downward speed v disappears from the top surface of the hourglass, and the same quantity (but different grains) appears in the bottom of the hourglass where it is brought to rest. Thus, we find an *upward* acceleration $a = v/dt$ acting on the mass dm, and hence a net force given as $\Delta F = dm \cdot a = v\frac{dm}{dt}$.

Obviously, the extra weight is extremely small in an actual hourglass—probably too small to observe. But Shen and Scott have confirmed the weight increase using a very sensitive scale, and an apparatus with a much larger mass flow rate than an hourglass. The version suggested here is much simpler than that of Shen and Scott, and yields results that do not require a scale having especially high sensitivity.

The apparatus consists of two small plastic soda or water bottles, whose plastic caps have been glued together so that one bottle remains upside down on top of the other with the caps on. Remove the caps from the bottles, and drill a 3/8-inch (1 cm)-diameter hole in the center of the caps, so that water can flow from one bottle to the other. However, you will find that if you were to fill one bottle with water, and place it the top position, the water flow is hindered by the air pressure in each bottle.

There are three ways to overcome this problem, one of which is impractical. One solution is to make a 1/8-inch (0.3 cm) hole in the side of each plastic bottle, about half way up. Now, if you fill one bottle half way (just up to the hole in the side to avoid a mess), and place it on top of the other, the water flows easily, since the pressure imbalance between the two bottles is continually corrected as the water flows. Another possible solution would be to use the "tornado tube" to connect the bottles, rather than drilling a hole in their sides and caps.[7] (The undesirable solution to facilitate water flow between the bottles would have been to use a hole larger than 3/8 inch in the caps joining the bottles, because in this case the water flow would not be smooth, since air must simultaneously flow up through the hole as the water flows down.)

To do the demo, fill one bottle half way, and place it in the top (inverted) position, keeping your fingers over the 1/8-inch holes in the sides of the bottles so as to prevent flow. Place the two bottles on a digital scale accurate to 0.1 gram, and observe the scale readings during the water flow between bottles, which lasts about 3 seconds. When I made the measurement, the scale reading during flow was observed to be 0.4 ± 0.1 grams greater than the scale reading when the flow stops. Given a flow lasting about 3 ± 0.5 seconds, and a water mass of 300 grams, we find the flow rate to be $\frac{dm}{dt} = \frac{300g}{3s} = 100$ grams/second. The velocity of the descending water surface was estimated as 2.5 cm/sec, which then yields for the predicted weight excess: $\Delta F = v\frac{dm}{dt} = 0.25 \pm 0.08$ grams times g, where the quoted uncertainty is based on the uncertainty in flow time. Thus, within the measurement uncertainties, the predicted and observed weight increases are in agreement.

Clearly, while the weight increase during flow is real, its value is a tiny fraction of the force of the descending water, so maybe the conventional wisdom about the weight being unaffected by the flow is not so terribly wrong. In fact, in an actual hourglass, where the time of flow is one hour rather than 3 seconds, both the mass flow rate and the velocity of the descending water (or sand) surface are reduced by a factor of 2,400 compared to our case. For an actual hourglass therefore, the predicted value of ΔF would be $\frac{1}{2,400^2}$ of our value, or about 43 billionths of a gram—assuming equal masses of water and sand.

2.13 Terminal velocity of falling coffee filters

Demonstration
By dropping paper coffee filters from various heights, you can measure their terminal velocity, and see how the force of air resistance or drag depends on velocity.

Equipment
Four coffee filters, two meter sticks, and a stopwatch.

Discussion

Any object after falling far enough reaches a terminal velocity, at which point the upward force of the air, or drag force, equals its weight. When dropped concave side up, coffee filters maintain their orientation as they fall, so that, unlike a dropped uncrumpled piece of paper, they are useful for studying air resistance, as suggested in a demo in *The Dick and Rae Physics Demo Notebook.*[8]

You can measure the average velocity of a coffee filter by timing its fall from various heights, say $h = 0.5, 1.0, 1.5, 2.0$ meters, and then computing $v = h/t$ for each height. The terminal velocity of a falling object is asymptotically approached, the greater the distance of fall, but a coffee filter is so light that you should find that within measurement uncertainties you get the same velocity for all four heights, which implies that the terminal velocity is reached long before the filter has fallen 0.5 meters.

Now, try dropping two filters—one on top of the other—side by side with one filter. The two filters, of course, have a higher terminal velocity, and will reach the ground first if dropped from the same height as one. See how much of a head start you need to give the one filter, so that it reaches the ground at the same instant as the two filters when they are released simultaneously. The ratio of the two heights for one and two filters, R, will be approximately the same as the ratio of the two terminal velocities, v_2/v_1, as long as we assume that the two filters combined attain their terminal velocity very quickly. Based on this height ratio (for simultaneous impacts of one and two filters), we can figure out how the drag force depends on velocity.

Let us assume that the drag force depends on some power of the velocity: $F_{drag} = Cv^N$, where C is a constant. You can use your observations with one and two filters to deduce the value of the exponent, N. To understand the method, remember that $F_{drag} = mg$ when the terminal velocity is reached, so two filters experience twice the drag force as one at their terminal speed. We can easily find the relation between the terminal velocities v_1 and v_2 when dropping one and two filters, using: $F_2/F_1 = 2 = v_2^N/v_1^N = R^N$.

The last equation can be solved for the exponent N to yield $N = \log 2/\log R$, For example, this relation shows that if the ratio of terminal velocities (or drop heights) for two filters and one filter is two, then the drag force is a linear function of velocity ($N = 1$), while if the ratio is $\sqrt{2}$ it would reveal the drag force to be a quadratic function of velocity ($N = 2$)—the latter being the result most observers find for low velocities in air. It is worthwhile to repeat the experiment by dropping three filters combined side by side with one, and then dropping four filters combined side by side with one—in both cases measuring the height ratios, R, needed to achieve simultaneous impacts.

Given these ratios, you could then see if you get a consistent exponent in the power law for the drag force when the terminal velocities of three and four filters are compared with one. The formula for comparing a drop of M filters with one filter is almost the same as what was found earlier, namely, the exponent in the force law is given by $N = \log M / \log R$. (If you don't find a consistent exponent with what you found from comparing two filters with one, within the limits of measurement uncertainty, this result might suggest one of two possibilities. Either the exponent N varies with velocity, so that the dependence of drag force on velocity is not strictly a power law, or else we need to reexamine the assumption that the terminal velocity is reached very quickly, so the ratio of terminal velocities is not equal to the ratio of drop heights.)

Notes

1. R. Ehrlich, *Turning the World Inside Out*, p. 4.
2. R. Ehrlich, The Physics Teacher, **32**, 51–53 (1994).
3. R. Ehrlich, *Turning the World Inside Out*, p. 34.
4. R. Chasnov and L. Overcast, The Physics Teacher, February 1990.
5. J. Walker, *The Flying Circus of Physics*, p. 26.
6. K. Y. Shen and B. L. Scott, The American Journal of Physics, **53**, 787–89 (1985).
7. Using an inexpensive "tornado tube" to connect the two bottles, you could avoid doing any construction whatsoever. When the connected bottles are then rapidly swirled prior to placement on the scale, the water smoothly flows in a vortex from one to the other.
8. D. R. Carpenter Jr. and R. B. Minnix, *The Dick and Rae Physics Demo Notebook*, p. M-136.

Chapter 3

Statics, Equilibrium, and Accelerometers

An object at rest or in equilibrium represents a special case of Newton's second law, where no acceleration is present. One demo included here (demo 3.3), is not, strictly speaking, an example of equilibrium, but rather the constant velocity motion resulting from a series of short accelerations and decelerations. The chapter also includes four demos for the design of various kinds of accelerometers. As with the demos in the previous chapter, those that are more challenging mathematically appear toward the end of the chapter.

3.1 Your back to the wall

Demonstration
With your back to the wall, and your feet flat on the floor with heels against the wall, you will find it extremely difficult to pick up an object on the floor in front of you.

Equipment
None.

Discussion
In order for you to remain standing and not to fall over, the center of mass of your body must remain within the base of support. The base of support includes the area under and between your feet. If your CM does not lie above this base area, gravity exerts a torque that causes you to fall over, unless you shift your feet. In fact, every time you take a step during walking, your body begins to fall, and the fall is stopped by your other foot.

  Suppose you have your back to the wall, with your heels on the floor and against the wall. Starting from this posture, any attempt to bend over to pick up an object on the floor will very likely shift the location of your CM forward far enough beyond the tips of your feet, and you will fall over. Without a wall behind you it is possible to bend over to pick something up because when you bend over from the waist your legs automatically angle backward so as to keep your CM directly over your feet.

There is another strategy you might think of trying to pick up something while your back is against the wall. Rather than bending from the waist, you might try bending your knees, and sliding down the wall. Unfortunately, if you try this method, you will also fail, because you will find it impossible to keep your heels on the floor. As you slide down the wall, your heels will invariably lift off the ground, because as your center of mass moves forward your weight gets shifted from your heels toward your toes.

A third possible strategy that actually might work if you have sufficient athletic prowess would be to allow the upper part of your body to fall forward stopping its fall with your hands, while your heels remain on the floor against the wall. After grabbing the object on the floor, you would then need to push the floor with your hands with enough force to drive your body back to its original vertical position. But such a push would not be easy to achieve. Too hard a push off the floor would cause you to bounce off the wall. Too easy a push—the more likely possibility—would not be enough to restore your body to the vertical position. One way to improve your chances of success would be to keep your feet as far apart as possible—approaching a split—before allowing your upper body to fall forward. In that way, you don't have to push off the floor quite as hard to restore your body to its vertical position. This method actually does work!

3.2 Avalanches in a sand pile

Demonstration
The avalanches occurring in a pile of sand can be understood in terms of the coefficients of static and sliding friction.

Equipment
A ziplock^{TM} bag half-filled with sand.

Statics, Equilibrium, and Accelerometers

Discussion

Sand and other granular materials have some properties they share with solids and other properties they share with liquids. For example, sand flows like a liquid, but it can stably maintain itself in a pile like a solid. The basic ideas considered here apply equally well to three-dimensional (conical) piles of sand as they do to two-dimensional piles (such as in a half-filled ziplock^{TM} bag), but obviously the latter case is more convenient to demonstrate.

Starting with the level of sand in the bag horizontal, slowly rotate the bag about a horizontal axis perpendicular to the bag. You should observe that when the sand surface makes a certain maximum angle θ_0 with the horizontal an avalanche occurs, after which the angle of the sand stabilizes at some smaller angle θ_1. The same behavior should repeat itself (at the same angles), if you continue rotating the bag.

We can understand the basis for the two angles θ_1 and θ_2 in terms of the different values of the static and sliding (kinetic) friction coefficients, μ_s and μ_k. It is easy to show that a mass on an inclined plane whose angle is varied will start to slide when the angle of the incline is given by $\tan \theta = \mu_s$, and that the mass slides at uniform velocity only if the angle is given by $\tan \theta = \mu_k$. The same rules apply to grains of sand in a sandpile.

For example, consider the behavior of the top layer of sand, as the slope of the incline of the sandpile is increased, as the bag is rotated. When the angle of the incline reaches $\tan \theta = \mu_s$ the sand will begin to slide, causing the shape and slope of the pile to change. When will the avalanche stop? Based on the definition of μ_k the sliding sand experiences a decelerating force (stopping the avalanche), when the angle of the incline satisfies $\tan \theta < \mu_k$.

3.3 Vibrating electric razor on an inclined plane

Demonstration

For a certain range of angles, a vibrating electric razor placed on an inclined plane will slide down the incline at a constant speed that depends on the angle of the incline.

Equipment

A board and a cordless electric razor with its case.

Discussion

Some electric razors have nonslip rubber bumps on them. If that is the case with yours, leave it in its case when you place it on the inclined plane. When the vibrating razor is placed at rest on the incline, you will probably observe the following behavior:

- For angles of the incline less than some value θ_0 the razor remains at rest.
- For angles greater than some value $\theta_1 > \theta_0$ the razor accelerates down the incline.
- For angles θ between θ_0 and θ_1 the razor moves down the incline at a constant velocity v that depends on θ.
- For angles only slightly above θ_0, the velocity v is extremely small, but it increases at an increasing rate as the angle θ approaches θ_1.

As we shall see, this interesting behavior of a razor on an incline can be explained using Newton's second law.

The primary effect of the razor's vibrations is to introduce a periodic variation in the normal force $N = mg\cos\theta$, exerted by the plane. Effectively, the constant g needs to be multiplied by the factor $1 + \epsilon\sin\omega t$, where ϵ is some small number, and ω is the frequency of the razor's vibrations, assumed to be normal to the plane. (If the razor were on a scale, for example, the scale reading would theoretically fluctuate rapidly about its average value, but almost certainly the damping of most scales would reduce such fluctuations to a negligible level.)

The force components acting along the incline include gravity, $-mg\sin\theta$ (down the incline), and friction, $f = \mu N = \mu mg\cos\theta(1 + \epsilon\sin\omega t)$ (up the incline). Thus, according to Newton's second law, the acceleration is given by:

$$a = \Sigma F/m = \mu g\cos\theta(1 + \epsilon\sin\omega t) - g\sin\theta. \qquad (3.1)$$

Let us consider for what angles of the incline the acceleration a defined by the preceding equation is either positive or negative for all values of time. If we required a to always be positive (up the incline), its smallest positive value clearly would occur at times when $\sin \omega t = -1$. Taking $a > 0$, and solving equation 3.1 yields the result that $\tan \theta < \tan \theta_0 \equiv \mu(1-\epsilon)$. Thus, for inclines having an angle less than θ_0, an initially moving razor will have an acceleration up the incline, and be decelerated to rest, while an initially stationary one will remain at rest.

Requiring a to always be negative (down the incline), can easily be shown to be equivalent to requiring that $\tan \theta > \tan \theta_1 \equiv \mu(1 + \epsilon)$. Thus, for inclines having an angle greater than θ_1, a razor will accelerate down the incline. The case where the incline angle lies in the range $\theta_0 < \theta < \theta_1$ is perhaps the most interesting. In this case, the acceleration of the razor is alternately positive and negative during successive portions of its 60-cycle vibrations.

Suppose a razor is placed at rest on an incline of angle θ in the interval $\theta_0 < \theta < \theta_1$. Initially the razor accelerates down the incline, and then is brought to rest during the positive acceleration phase of each vibration, so the razor literally takes a series of short steps down the incline—one for each vibration—resulting in a uniform velocity down the incline. Obviously, the velocity magnitude depends on the fraction of the cycle, f, that the acceleration is positive. Clearly, we must have $f = 0$ for $\theta = \theta_0$, and $f = 1.0$ for $\theta = \theta_1$, based on the earlier definitions of these angles. Thus, we expect the (constant) velocity of the razor to increase as θ is increased from θ_0 to θ_1. In fact, the velocity increases quite rapidly, because not only does f increase as we go from θ_0 to θ_1, but so does the magnitude of the negative acceleration down the incline.

The type of motion discussed in this demo—continual starts and stops resulting in an average uniform speed—is quite analogous to the "drift speed" of electrons in a wire with an applied voltage. In both cases the effect of the start-stop motion is to create an average uniform *velocity*, not an acceleration which we would normally associate with an outside force. If this demo were used to model the drift velocity of electrons in a wire, the slope of the incline corresponds to the voltage across the ends of the wire.

3.4 A simple accelerometer for use on the OHP

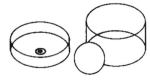

Demonstration
An accelerometer made from a steel ball resting in the hole of a small washer can be used to measure accelerations above some threshold value, as the device is moved around on an overhead projector in several types of motions.

Equipment
A stainless steel ball of diameter one inch (2.54 cm), a small metal washer having a 1/8-inch (0.3 cm)-diameter hole, some plastic cement, and a cylindrical transparent plastic box having a diameter of about 5 centimeters. Such plastic boxes can be obtained at scientific equipment companies, pharmacies, or optometrists. You only need the cover of the plastic box to make the device.

Discussion
Constructing your accelerometer is extremely simple. First, invert the cover to the plastic box so that its flat side rests on a table. Then glue the small washer to the center of the cover. Be sure that glue does not fill up the washer hole. After the glue dries, place the steel ball, so that it sits on top of the washer hole. Your accelerometer is complete! Place the device on a horizontal surface such as an overhead projector and move it around on the surface. The ball will remain on top of the washer hole only if the acceleration of the device is less than some threshold value. We can predict the threshold acceleration using Newton's second law, and the concept of "apparent" inertial forces (such as centrifugal force), which act in noninertial reference frames.

Suppose the device is being accelerated with an acceleration a. If the ball is just on the verge of popping out of the washer hole, only one side of the washer hole exerts any force on the ball. Therefore, the normal force vector exerted by the washer on the ball extends from a point on the washer hole in the direction of the ball's center, and makes an angle with the vertical given by $\sin \theta = d/D$, where d is the diameter of the washer hole, and D is the diameter of the ball. But, since the y-component of this normal force counteracts gravity ($N_y = mg$), and its x-component provides the net force to accelerate the ball ($N_x = ma$), the angle it makes with the vertical must satisfy $\tan \theta = a/g$. In the small angle approximation, we therefore obtain the final result that the ball is on the verge of coming out of the hole when its acceleration satisfies $a = g \tan \theta \approx g \sin \theta = d/D$. Given a ball of diameter D eight times that of the washer hole d, this last result implies that the device is "triggered" at a horizontal acceleration $a = g/8$, that is, one-eighth of a "g."

A simple calibration check of the device would be to put it on a board and gradually increase the angle of the incline to the point where the ball just rolls off the washer. Tilting the incline by an angle θ with the horizontal is equivalent to accelerating the device horizontally by an acceleration $a = g \tan \theta$, so the maximum angle of the incline before the ball rolls off the washer should be given by $\tan \theta = d/D$, or 0.125 in our case. (This calibration method is much easier than trying to achieve a constant horizontal acceleration of $a = g/8$.)[1]

When using the device as an accelerometer on the overhead projector, you might want to try two types of motions: uniform circular motion (UCM), and simple harmonic motion (SHM), both of which you can approximate by hand motions. The acceleration in the case of UCM can be expressed as $a = \omega^2 r = 4\pi^2 r/T^2$, where r is the radius of the circle, and T is the time for one revolution—or the period. One easy way to move the device in a circular path is to move it in contact with the inside wall of a 12-inch (30 cm)-diameter embroidery hoop.

You could try to keep the time for one revolution equal to some number of seconds by using a metronome or by having someone count off the seconds as you move the device, and then try shorter and shorter periods until the device is just triggered. From that shortest period and the known radius, you could compute the observed threshold acceleration, and compare with the predicted value. It is not so easy, however, to move the device at a precisely constant speed manually. Fluctuations in the speed will almost certainly cause the device to be triggered when the acceleration computed from the preceding formula is significantly less than the predicted threshold value. Specifically, the measured threshold acceleration found from $a = 4\pi^2 r/T^2$ is likely to be no more than two-thirds the value predicted from $a = gd/D$ because of these fluctuations.

You may find it somewhat easier moving the device back and forth in an approximation of simple harmonic motion than uniform circular motion. For SHM the acceleration is given by $a = \omega^2 x = 4\pi^2 x/T^2$, where T is the period, and x is the displacement from equilibrium. Obviously, the acceleration is greatest when x is the maximum displacement from equilibrium—at the *turning points* of the motion. One way to do the experiment for SHM would be to move the device back and forth at a fixed period (matching that of a metronome), with a larger and larger amplitude, and see at how large an amplitude it is triggered. Again, due to nonuniformities in your hand motion, it is likely to be triggered at an acceleration significantly below the predicted threshold value. (This demo is based on an article I wrote for The Physics Teacher.)[2]

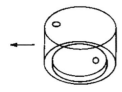

3.5 A second accelerometer for the OHP

Demonstration
Another accelerometer, made from a concave lens on which a small steel ball rolls, can be used to measure accelerations (based on the position of the ball on the lens), as the device is moved around on an overhead projector in several types of motions.

Equipment
A stainless steel ball of diameter 1/8 inch or 0.3 cm (a BB), a concave lens, baby oil, plastic cement, and a cylindrical transparent plastic box having a diameter of about 5 cm. As noted in the previous demo, such plastic boxes can be obtained at scientific equipment companies, pharmacies, or optometrists. The concave lens should be plano-concave, that is, flat on one side, and should have a power of -3.0 diopters, or a focal length $f = -1/3$ meters. The lens is not being used as an optical device here, only as a surface on which to allow the small ball to roll. The lens diameter should be small enough so that when the lens is glued to the center of the bottom interior of the box (flat side of lens down), there is enough space between the edge

of the lens and the walls of the box for the small ball to roll off the lens and into the gap.

Discussion

To construct this version of the accelerometer, glue the lens to the center of the interior bottom of the plastic box. Place the small ball on the lens, and notice that when you gently accelerate the device the ball oscillates. If the acceleration is sizable enough, the ball leaves the surface of the lens, and falls into the gap between the lens and the sides of the box. In order to dampen the oscillations of the ball as it rolls on the lens, fill the cylindrical box with baby oil, and put the cover on.

Now when you accelerate the device gently, the ball should move backward (opposite to the acceleration direction), and exhibit much less oscillation. In fact, it can be shown that, under certain conditions, the amount the ball moves backward is directly proportional to the acceleration of the device. One added benefit of using the baby oil is that you are likely to trap a small air bubble when putting the cover on the box. As a result, you will find that when the device is accelerated, the air bubble and steel ball move in opposite directions, and both can be clearly seen on the overhead projector.

In the previous demonstration (the ball-washer accelerometer), we had a device that only showed when the magnitude of the acceleration exceeded some value. In contrast, the present device shows both the magnitude and direction of the acceleration, based on the direction the ball rolls. In addition, this device gives some visible indication when it is about to be "triggered," that is, when the ball is close to rolling off the edge of the lens into the gap. (Incidentally, you can, of course, easily get the ball back on the lens by properly shaking the oil-filled box.)

You might try using the device in the same two kinds of motions as suggested in the previous demonstration—uniform circular motion and simple harmonic motion. The predicted acceleration needed for the ball to roll off the lens can easily be shown to be $a = gr/R$, where r is the radius of the lens, and R is its radius of *curvature*. The radius of curvature of the plano-concave lens can be easily found in terms of its focal length, f, and index of refraction, n, using: $R = (n - 1)f$. So you could see whether the measured acceleration for the ball to roll off the lens, given by $a = gr/R = gr/[(n - 1)f]$, is close to the predicted value when the device moves in UCM or SHM. As shown in the previous demo, for example, for SHM the acceleration is predicted to be $a = \omega^2 x = 4\pi^2 x/T^2$, where T is the period, and x is the displacement from equilibrium. However, as noted in the previous demo, the measured acceleration is likely to be no more than two-thirds that predicted, because of nonuniformities in the motions when done manually. (This demo is based on an article I wrote for The Physics Teacher.)[3]

3.6 A vibrating ruler accelerometer

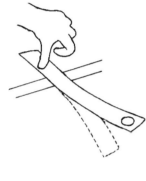

Demonstration

If you pluck the end of a ruler hanging over a table like a diving board, a penny placed at the end of the ruler signals that its acceleration exceeds one g when the penny can be seen and heard to clatter.

Equipment

A small C-clamp, a plastic 12-inch (30 cm) ruler, and some pennies. It would also be useful to use a strobe to measure the frequency of the vibrating ruler, but it is not essential.

Discussion

An object undergoing simple harmonic motion of amplitude x should have a maximum acceleration given by $a = \omega^2 x = (2\pi f)^2 x$. That maximum acceleration will occur at the turning points of the motion, where the displacement from equilibrium is maximum. For a sufficiently large amplitude, the acceleration must exceed g, which occurs when $x > g/(2\pi f)^2$. You can verify the preceding relation using a vibrating ruler. Clamp one end of a horizontal ruler to a table with its grooved side down (or else just hold the end of the ruler on the table firmly in place by hand). You may find it convenient to decrease the ruler's vibration frequency by taping about six pennies to its underside near its free end. (The added weight will make the amplitude x at which $a = g$ larger, and easier to measure.)

Place one free penny on top of the ruler at its end. Pluck the ruler, and observe the penny as the ruler oscillates. You should find that for small oscillations the penny remains in contact with the ruler at all times, but for large oscillations it loses contact, causing an audible clatter. It may even jump off the ruler in some cases. Measure the largest vibration amplitude for which the penny remains in contact with the ruler at all times (no clatter). The penny should begin to lose contact with the ruler when the ruler's downward acceleration just exceeds g, so the ruler is literally falling out from underneath the penny. See if the amplitude at which this occurs is given by: $x = g/(2\pi f)^2$, where f is the vibration frequency, measured by a strobe, or by timing the oscillations.

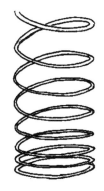

3.7 Static equilibrium of a suspended slinky

Demonstration
The spacing between successive turns of a slinky suspended vertically under its own weight can be used to test Hooke's law.

Equipment
A slinky and a meter stick.

Discussion
Slinkys are wonderful toys, useful in all kinds of demos. If you hang a slinky from some point, so that its bottom end is just above the floor, it will stretch under its own weight. As you might expect, the spacing between turns is greater the higher up you look, because the higher a turn is, the greater the weight it must support below it. As shown in an article by A. P. French,[4] on which this demo is based, the spacing between adjacent turns is in fact a linear function of n, the integer identifying a particular turn. French shows that the spacing Δz between turns number n and $n + \Delta n$ is given by the relation:

$$\Delta z = \Delta n \left[\frac{L_0}{N} + \frac{2}{N^2}(L - L_0)\left(n + \frac{\Delta n}{2}\right)\right]. \qquad (3.2)$$

In the preceding relation L is the length of the slinky when it hangs freely, L_0 is its unstretched length, and N is the total number of turns suspended. The key assumption in the derivation of the preceding formula is Hooke's law—a linear relation between the amount of stretch and the stretching force. In fact, without even going through the derivation, it is clear that Δz should increase linearly with turn number n, because for each increase of n by one (or by Δn), the amount of additional weight is constant, so the extra stretch (or increase in Δz), should be constant if Hooke's law is obeyed.

In suspending the slinky vertically, you may find it easier not to suspend it from its very last turn, but instead clamp some turns at the end together, so as to give it structural strength at the point of suspension. When you make measurements on the spacing between turns, you probably won't want to bother measuring the spacing between adjacent turns ($\Delta n = 1$), but instead you might use every fifth or tenth turn ($\Delta n = 5$ or 10). If you plot your experimentally measured spacings Δz versus turn number, n, you are likely to find extremely close agreement with equation 3.2—showing that the slinky satisfies Hooke's law quite closely. In fact, as noted by French, the agreement he found seems almost better than one would expect, given that most springs satisfy Hooke's law only approximately.

3.8 A row of magnetic marbles on an incline

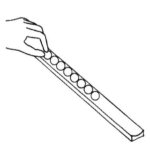

Demonstration
If you position a row of magnetic marbles on an inclined plane, while holding the topmost marble fixed, the number of marbles that can remain attached magnetically depends on the angle of the incline.

Equipment
A collection of magnetic marbles, which can be obtained from scientific equipment companies or toy stores. For an inclined "plane," you may find it useful to use a length of concave molding, because the concavity is useful in keeping the marbles from rolling off the sides of the incline.

Discussion
Form a row of magnetic marbles on a piece of molding whose angle can be varied. You will find that the marbles automatically align themselves properly

if you build up the row one marble at a time on the piece of molding. You can easily measure the coefficient of static friction μ between the row of marbles and the molding surface by varying the angle of the incline, and observing at what angle θ_0 the row of marbles just begins to slide down. The coefficient of friction is then found using $\mu = \tan \theta_0$.

Now suppose you hold the topmost marble in a row fixed, while you increase the angle of the incline past the value θ_0. The marbles do not slide down, provided the combined forces of friction and the magnetic attraction up the incline exceed the component of gravity down the incline. In its qualitative version of this demo, point out that the angle of the incline at which the marbles break free from the one you hold down decreases as the number of marbles in a row increases. However, we can be much more quantitative, and predict the angles for any number of marbles in a row.

In particular, in a row of N marbles *just on the verge* of sliding down the incline (with the topmost one held fixed), the forces on the $N - 1$ marbles below the topmost marble include: (1) the magnetic attraction of the top marble for the one next to it, T; (2) static friction up the plane, given by $(N - 1)\mu mg \cos \theta$; and (3) the component of gravity down the incline, given by $(N - 1)mg \sin \theta$. Thus, we have: $T + mg(N - 1)(\mu \cos \theta - \sin \theta) = 0$, or

$$T = mg(N - 1)(\sin \theta - \mu \cos \theta). \tag{3.3}$$

Notice that this same equation can be applied to any marble in the row—say, marble k—so that, provided the magnetic forces between marbles are all equal strength, you should find that the row of marbles will break at the topmost marble, since the right-hand side of equation 3.3 is greatest when $N = k$. If you don't find this to be true, remove any "weak link" marbles, and try again.

Using the identity $\cos \theta = \sqrt{1 - \sin^2 \theta}$, equation 3.3 can easily be solved for $\sin \theta$, yielding

$$\sin \theta = \frac{2\alpha + \sqrt{4\alpha^2 + 4(1 + \mu^2)(\mu^2 - \alpha^2)}}{2(1 + \mu^2)}, \tag{3.4}$$

where $\alpha = T/mg(N - 1)$.

Now try the following experiment: for various numbers of marbles in a row—say, between 10 and 20—while you hold down the topmost marble, see what is the maximum angle of the incline before the marbles below the topmost marble breaks free from it. (Be sure to always use the same pair of marbles for the topmost two as you vary the number of marbles.) This experiment will give you different θ values for each number of marbles in a row. Actually, instead of θ, you could simply tabulate the vertical height of the top of the incline, which is related to θ through $y = L \sin \theta$, where L is the length of the incline.

The table shows the results of one such experiment. The second column of the table shows the heights y in centimeters measured for the top of the incline when using numbers of marbles listed in the first column. The third column shows the predicted heights from equations 3.3 and 3.4, which show excellent agreement. Here is the exact procedure for finding the predicted heights:

- Find μ from $\mu = \tan \theta_0$, where θ_0 is the smallest angle with the horizontal at which the row of marbles starts to slide down the incline starting from rest, without holding any of them.
- Use equation 3.3 to find T using one measured height y and the associated value of $\sin \theta = y/L$. (For the tabulated measurements, the height for 16 marbles was used.)
- Use equation 3.4 and $y = L \sin \theta$ to find all the remaining "predicted" heights. These are the values shown in column 3 of the table. (As a check you should also calculate the height for the one case that you used to determine T, just to be sure the predicted and measured height agree for this one.)

N	y_{meas}	y_{pred}
10	141	147
12	133	131
14	122	120
16	112	112
18	107	106
20	100	100

Statics, Equilibrium, and Accelerometers

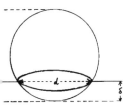

3.9 A 1000 g accelerometer

Demonstration

By dropping steel balls onto a wooden board, you can determine their acceleration during impact by measuring the diameter or depth of their dents, and comparing the results with predictions.

Equipment

A board made of soft wood, and a stainless steel ball one inch (2.54 cm) in diameter.

Discussion

To "calibrate" the 1000 g accelerometer, you need to see what depth dents in the board are produced by a known force on the ball. This calibration can be done if you place the steel ball on the board, and step on the ball with all your weight resting on that foot—but don't jump on it. Try to have all your weight on the ball itself with none of it on the board directly. Since the ball is pressing on the board with both your weight (Mg) and its own weight (mg), the total force on the board is $(M + m)g$, which is $\frac{M}{m} + 1$ times the weight of the ball alone, or $(\frac{M}{m} + 1)$ g's of force—a number in the vicinity of 1000 g's. After producing several dents this way, measure their average diameter, d, and also the diameter of the ball itself, D. A more useful quantity for finding the impact force is the average depth of the dents, not their diameter, but the latter is much easier to measure, and you can compute the depth from the diameter using the geometrical relation: $\delta = \frac{4D}{d^2}$.

It is surprisingly easy to predict the initial height from which a ball needs to be dropped to produce a dent having the same diameter (and depth) as you found when stepping on it. When the ball is in free fall with acceleration g through a distance h, it acquires a velocity v just before impact given by $v = \sqrt{2gh}$. If it is now decelerated by the impact from that same velocity v down to zero in a distance δ, the deceleration required is simply $a = v^2/(2\delta) = gh/\delta$, which can be solved for h, giving $h = a\delta/g$.

As we showed earlier, the dents produced by stepping on the ball correspond to $(\frac{M}{m} + 1)$ g's, so substituting the acceleration $a = (\frac{M}{m} + 1)g$ in the preceding equation for h gives $h = (\frac{M}{m} + 1)\delta$. Finally, we can use an earlier relation between dent depth δ and diameter d (namely, $\delta = \frac{4D}{d^2}$), to obtain

$$h = \frac{4D}{d^2}\left(\frac{M}{m} + 1\right). \tag{3.5}$$

Equation 3.5 allows us to predict the drop height h which will produce dents having the diameter d found when a person of mass M steps on a ball of mass m and diameter D. Try dropping the ball from a variety of heights, and see how close you come to the prediction.

3.10 High friction Atwood's machine

Demonstration
Make an Atwood's machine by hanging two weights connected to the ends of strings draped over a smooth cylinder, and use the device to measure the predicted *exponential* increase in friction exerted by the cylinder, as the string is wound 1, 2, 3, or more turns around the cylinder.

Equipment
A collection of known masses, some string, and a smooth cylinder, such as a pen.

Discussion
Normally, in an Atwood's machine we strive to keep friction as low as possible by passing the string over a pulley. Here, however, we want to use a fixed cylinder (such as a pen) instead of a pulley, in order to see how friction depends on the number of turns wound around a cylinder. Drape the string over the cylinder, and hang a mass m_1 on one side. On the other side hang enough mass $m_2 > m_1$, so that the heavier mass just overcomes friction, and begins to slide downward. Define the ratio $m_2/m_1 = R_0$, when the heavier mass just begins to slide downward, pulling the lighter mass upward. Now, repeat the experiment, only this time wind an extra loop of string over the cylinder. In doing so, be sure that the string does not lie on top of itself, but is in direct contact with the cylinder itself. Also, you need to slowly release your support of the heavier mass, so that the string has a chance to tighten uniformly around the cylinder.

Obviously, the friction in this case will be much more than before, and therefore the value of the mass ratio m_2/m_1 (defined as R_1), for which the masses just begin to slide will be greater than before. You might repeat the experiment a few more times, each time winding the string one extra turn over the cylinder, and observing the value of the two masses when they are just able to slide. You may want to keep the value of m_1 constant as m_2 takes on progressively greater values.

Suppose that you measure mass ratios for the string wound 0, 1, 2, 3, and 4 turns over the cylinder. Let us define R_k as the ratio of the masses just able to overcome friction when k turns of string are wound over the cylinder. As shown later, the measured mass ratios R_0, R_1, R_2, and R_3 are predicted to have a simple relationship: $R_0 = R_1^{1/3} = R_2^{1/5} = R_3^{1/7} = R_4^{1/9}$. When I did the experiment using $m_1 = 20$ grams, the m_2 values for 0, 1, 2, 3, and 4 turns were found to be, respectively: 35, 90, 270, 500, and 1500 grams, which yields for: R_0, $R_1^{1/3}$, $R_2^{1/5}$, $R_3^{1/7}$, $R_4^{1/9}$ the values 1.75, 1.65, 1.68, 1.58, and 1.62, showing them to be reasonably constant, as predicted. The proof of the preceding relationships between the mass ratios is a little lengthy, and will take the rest of this discussion.

Consider a string wound around a cylinder through a small angle $d\theta$. If the force at one end of the string is f, the force at the other end will be $f + df$, where the difference is due to the friction force between the cylinder and the string. The friction force may be expressed as μN—if the string is just on the verge of sliding—where N is the normal force and μ is the friction coefficient.

In equilibrium, the four forces, f, $f + df$, N, and μN must have a zero resultant. Therefore, from the force diagram, we see that $df = \mu N$ and $N = f d\theta$. Combining these two relations yields $\frac{df}{f} = \mu d\theta$, which can be integrated to give $f = f_0 e^{\mu \theta}$. Finally, if the angle θ in radians is converted to a number of turns n, namely, $\theta = 2\pi \mu n$, the previous relation becomes

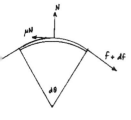

$$\frac{f}{f_0} = e^{2\pi \mu n}$$

where n is the number of turns the string is in contact with the cylinder, μ is the coefficient of friction, and f_0 and f are the forces on the two ends of the string.

In the case of forces exerted by two hanging masses, we have $f = m_2 g$, and $f_0 = m_1 g$. Thus, applying the preceding equation to the simple Atwood machine (string in contact with the cylinder for n = 1/2 turns), yields the result

$$\frac{f}{f_0} = \frac{m_2}{m_1} = R_0 = e^{\pi \mu}.$$

For the case where the string has 1, 2, 3, or 4 extra turns, we need to use $n = \frac{3}{2}, \frac{5}{2}, \frac{7}{2}, \frac{9}{2}$, giving the results $R_1 = e^{3\pi\mu}$, $R_2 = e^{5\pi\mu}$, $R_3 = e^{7\pi\mu}$, $R_4 = e^{9\pi\mu}$. Finally, we can see that $R_0 = R_1^{1/3} = R_2^{1/5} = R_3^{1/7} = R_4^{1/9}$, as claimed earlier.

3.11 Ladder against the wall

Demonstration
The maximum lean angle of a ladder against a wall depends on the coefficient(s) of friction and the location of a person on the ladder.

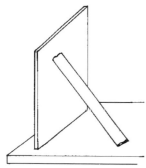

Equipment
A 12-inch (30 cm) plastic ruler (the "ladder"), and some clay (the "person"). A second ruler will be needed to measure the distance of the height of the top of the ladder on the "wall," which could be a vertical note pad, for example.

Discussion
This standard problem in elementary physics texts is usually made solvable by assuming there is no friction between the ladder and the wall. Some such assumption is necessary, because otherwise there are four unknowns (the normal and friction forces at each end of the ladder), but only three equations (the vanishing sum of the torques, and the horizontal and vertical force components).[5]

For simplicity, however, let us suppose that we can assume there is no friction at the wall, and that the ladder of length L and mass M makes an angle with the vertical θ just before slipping. Let us further assume that a person of mass m has climbed up the ladder a maximum distance x from the bottom just before it slips. The normal force at the base of the ladder must just balance the combined weight of the

person and ladder $N = (m+M)g$, so the friction force
at the floor is therefore $f = \mu N = \mu(m + M)g$. The
only other horizontal force component in the problem is the normal force of the wall on the ladder, so
the normal force of the wall must have the same magnitude (but opposite direction) as the friction force
of the floor. Setting the sum of the torques about the
bottom of the ladder to zero yields the equation

$$\mu(m + M)gL \cos \theta - Mg\frac{L}{2} \sin \theta - mgx \sin \theta = 0. \quad (3.6)$$

In the special case of $m = 0$ (no one on the ladder),
the preceding equation yields $\mu = \frac{1}{2} \tan \theta_0$, where the
subscript reminds us that this maximum angle with
the vertical holds for the case $m = 0$. Thus, one way to
find the coefficient of friction between the base of the
ladder-ruler and the floor would be to find its maximum lean angle with the wall without any additional
weight on it. This can be done most simply by measuring the vertical position y of the top of the ruler on
the notepad-wall just when slipping begins, and find-
ing θ from the relation $\theta = \cos^{-1}(y/L)$, rather than
using a protractor.

Equation 3.6 can be solved for the angle θ to yield

$$\tan \theta = \frac{2\mu L(m + M)}{2mx + ML}. \quad (3.7)$$

You could try to verify the correctness of equation 3.7
by adding mass m to the ruler (in the form of a lump
of clay) at a location a distance x from the bottom,
and see how closely the maximum angle before slipping agrees with the prediction. Of course, before
you can calculate the predicted angle with the clay
from equation 3.7, you need to first measure μ by
observing the maximum angle θ of the ladder without
the clay, and using the previously mentioned relation
$\mu = \frac{1}{2} \tan \theta_0$.

Variation using a real ladder
If you prefer that your demos have added realism,
you might consider using a real ladder in this demo,
if you have one handy. However, it should be a ladder
of uniform width along its length. It might be useful if
you added some wheels at the top of the ladder, and
perhaps added a pair of sneakers at the bottom to

minimize wall friction and maximize floor friction. If you do the demo using a real ladder, you could add some drama by predicting the fraction of its length ($f = x/L$) you could climb up before the ladder slips, rather than the maximum angle with the vertical for a fixed value of f.

On the other hand, the thought of being halfway up a ladder as it begins to slip may not appeal to you—in which case you may wish to use a "stand-in" in the form of a heavy weight that could be hung from successively higher rungs. We can use equation 3.7 to predict what angle with the vertical, θ, you should set the ladder at in order that it not slip until a mass m is a fraction $f = x/L$ up the ladder. If we substitute $\mu = \frac{1}{2}\tan\theta_0$, $f = x/L$ in equation 3.7, and write the ratio of the mass of the ladder to that of the added weight as $R = M/m$, the equation can be rewritten as

$$\tan\theta = \left(\frac{1+R}{2f+R}\right)\tan\theta_0. \tag{3.8}$$

Recall that θ_0 is the maximum angle the ladder can make with the vertical before slipping with no added mass, and θ is its actual angle with the vertical. For the special case where $f = 1/2$, we find that $\theta = \theta_0$. for any mass ratio, R. (Do you see why that should be obvious?) If you want to be able to reach the 3/4 point before slipping occurs just use $f = 3/4$ and the measured mass ratio R in equation 3.8 to find the appropriate angle of the ladder. You could then make a very suspenseful test of the prediction by placing the added mass at higher and higher rungs.

Notes

1. If you do try to calibrate the device by accelerating it uniformly, be sure that the acceleration is in the horizontal direction. For example, accelerating it down an inclined air track would not work; rather, you would need to accelerate an air track cart using a weight and pulley system.
2. R. Ehrlich, The Physics Teacher, article in press.
3. Ibid.
4. A. P. French, The Physics Teacher, **32**, 244–45 (1994).
5. Notice that we cannot assume the relations $f = \mu N$ hold at either the floor or wall, because in general we don't

Statics, Equilibrium, and Accelerometers

know where slipping first begins. In the general situation of a rough wall, discussed by Kenneth Mendelson, the equilibrium is determined by how the ladder is placed against the wall, and its elasticity. See K. Mendelson, The American Journal of Physics, **63**, 148–50 (1995).

Chapter 4

Orbital Motion and Angular Momentum

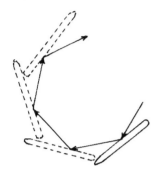

4.1 Effect of many sideways impacts

Demonstration
A ball moving on a flat horizontal surface such as an overhead projector can be made to move in a path approximating a circular orbit by delivering many impacts at right angles to its direction of motion.

Equipment
A one-inch (2.54 cm)-diameter steel ball and a pen.

Discussion
Place the ball on a horizontal surface such as the OHP, and be sure the surface is level, so the ball does not accelerate in any direction when released. Now start the ball rolling and deliver many rapid impacts at right angles to its path using the *side* of your pen. A good way to do this is by keeping the pen oriented tangent to the ball's curving path, and continually moving it ahead of the ball. The virtue of using the side of your pen is that you will find it easier to keep the ball in a closed (polygonal) path.

4.2 Tangential speed at the top of a wheel

Demonstration
The tangential speed of a point on top of a rolling wheel is twice that of the wheel itself, as can be easily seen using a ruler placed on a soda can on the overhead projector.

Equipment
A ruler and a soda can.

Discussion
Hold one end of the horizontal ruler, and rest the other end on the side of the soda can, which is on a horizontal surface such as the OHP. Move the ruler over the can thereby propelling the rolling soda can

forward, because of slight downward pressure you exert on it from the ruler. Observe that as it rolls, the can moves forward only half the distance of the ruler itself. How to explain this behavior? The ruler, of course, moves at the same speed as a point on top of the can if there is no slipping between ruler and can. Since the contact line between the can and the surface on which it rolls is momentarily at rest, and since this line is also the momentary axis of rotation, a point on top of the can has a tangential velocity $v = \omega r$, where $r = 2R$, and R is the can's radius. This velocity is exactly twice the velocity of the can's center of mass.

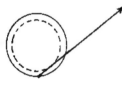

4.3 Pulling a spool with a thread

Demonstration
A spool can be made to roll either backward or forward when you pull the thread wrapped around it, depending on the angle of the thread with the horizontal.

Equipment
A spool of thread and a ribbon.

Discussion
To make the spool roll forward you need to pull the thread at a very small angle with the horizontal, while to make it roll backward you need to pull it at a large angle. The critical angle separating the forward and backward rolling motions, shown in the figure, is defined by extending the line of the pulled thread so that this line passes through the point of contact between the spool and the table. A force directed along this line produces zero torque on the spool about the contact point. Ideally, the spool should therefore not roll, and merely rotate in place as you pull the thread at this critical angle. Note that unless the thread comes off the spool at its center, the spool orientation is not stable when you pull the thread at a very shallow angle. You may get better results if you wind a ribbon around the spool, rather than a thread.

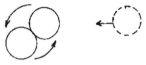

4.4 Colliding magnetic marbles

Demonstration
When you roll one magnetic marble toward another
to make a glancing collision, the two marbles rapidly
rotate about their CM when they stick together.

Equipment
Two magnetic marbles, which can be obtained at toy
stores or scientific equipment companies.

Discussion
It is commonplace to think of angular momentum as
a quantity possessed only by rotating objects, and not
linearly moving ones. But, we must not forget that
a linearly moving mass does, in fact, possess angu-
lar momentum $\vec{\ell} = \vec{r} \times \vec{p}$, where $\vec{p}$ is the linear momentum and $\vec{r}$ is the position vector relative to some
axis. You can easily show that linearly moving objects possess angular momentum by projecting one
magnetic marble into a second stationary one on an
OHP. When the marbles collide and stick together,
they rapidly rotate about their CM. The rapidity of
the rotation is greater the more glancing the collision,
and for head-on collisions no rotation is observed.

One *incorrect* interpretation of this demo is that
the initial linear momentum of the marble is "converted" to angular momentum during impact. It
might be worthwhile to explicitly raise this misinterpretation with the audience in order to consider
why it is incorrect. (The angular momentum existed
before the impact, as discussed above, and the linear momentum quickly disappears after impact due
to the force of friction.) You should be able to dis-
pel this misinterpretation by pointing out that the
marbles, in fact, can be seen to have some linear momentum immediately after the collision until friction
brings their center of mass to rest.

4.5 A precessing orbit

Demonstration
The elliptical orbit of a ball rolling on a concave surface will precess in a predictable manner.

**Orbital Motion and
Angular Momentum**

Equipment

A quarter-inch (0.6 cm)-diameter steel ball, and a transparent plastic hemispherical dome. Such domes may be purchased in crafts stores or from scientific equipment companies. A large one (diameter around 18 inches or 50 cm) would be preferable to a small one.

Discussion

Roll the ball on the concave surface of the dome resting on the OHP. Depending on the ball's initial velocity magnitude and direction, you should be able to achieve orbits that are either circular, linear, or elliptical. However, you should find that the elliptical orbits do not quite close, but instead they slowly precess in the direction of the ball's rotation. It can be shown that the precession rate, or the rotation of the major axis of the ellipse, is given by $\alpha = 3\pi ab/(16r^2)$, where a and b are the lengths of the major and minor axes of the ellipse, r is the radius of curvature of the dome, and α is the rotation angle in *radians* that the ellipse rotates each time the ball completes an orbit. The derivation of this formula is quite messy, and involves the "second approximation to the spherical pendulum."[1]

To make a quantitative test of the formula, you will need to measure the precession rate experimentally. In order to estimate the values of a and b for a given orbit, you should place a piece of transparency made from graph paper underneath the inverted dome, and watch the orbits on the OHP screen. It might be easiest if someone else simultaneously estimates how large an angle the ellipse precesses each orbit. If you just want to show that the observed results are in qualitative agreement with the formula, note that it predicts: (1) zero precession for a straight line orbit, for which $b = 0$; (2) greater precession for ellipses having a given size major axis, a, if their minor axis is of comparable size—that is, the more eccentric the ellipse, the less it precesses; and (3) greater precession for larger ellipses rather than small ones.

The mathematics used in deriving the precession of the orbit is identical to that used to find the precession of the orbit of Mercury—a key historical result supporting general relativity. However, it might be somewhat misleading for several reasons if you were to use this demonstration as an analogy to orbital precession in general relativity—mainly because the orbits here are not the result of the gravitational deformation of space by some central mass, but also because the orbits here result from a deformed surface whose shape does not remotely correspond to the gravitational $1/r$ variation in potential. For a somewhat less misleading demonstration of orbital precession in general relativity, see *Turning the World Inside Out.*[2]

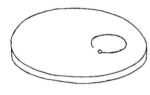

4.6 Ball on a rotating turntable

Demonstration
A ball released on a rotating turntable exhibits a surprising range of orbits, depending on its initial conditions.

Equipment
A turntable, such as a record player, and a solid ball, such as a one-inch (2.54 cm)-diameter steel ball.

Discussion
Be sure the turntable is accurately levelled by checking that the ball does not roll off when the turntable is at rest. While the turntable is spinning, release the ball at some point on the turntable, and it will probably fly off—as we might expect. Try releasing the ball again, only this time hold the ball gently in place on the spinning turntable using a folded index card, so as to allow it to spin up to speed before releasing it. "Spinning up to speed," means that when the ball is released, there is no slipping between the spinning ball and the turntable. The no-slipping condition requires that the ball's rotational speed along an axis pointing toward the center of the turntable is given by $\omega = v/r$, where v is the tangential speed of a point on the turntable at the distance r from the center where the ball is located.

When the ball is released from rest after spinning up to speed, its center of mass will remain nearly at rest on the spinning turntable—although when observed over a long time, it will be found to slowly approach the turntable center. An even more remarkable behavior occurs if the ball is given a slight push rather than released from rest, after it is allowed to spin up to speed. In this case, you will find that the ball travels in a circular orbit not necessarily concentric with the axis of the turntable. According to theory, it should complete an orbit in a time which is 3.5 times the period of rotation of the turntable, independent of the size of the orbit—an easily testable prediction.

Finally, modify the setup so as to give the turntable a very slight incline prior to releasing the ball. This time the ball's circular orbits will be found to slowly migrate in a direction perpendicular to the incline. A mathematical description of the orbits is identical to that in the case of a charged particle moving in crossed electric and magnetic fields. A full discussion of the theory behind this interesting phenomenon, and a comparison with experiment, can be found elsewhere.[3]

4.7 Hero's engine made from a soda can

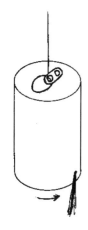

Demonstration
A water-filled soda can hanging from a string begins to spin as water streams out of a hole in its side, provided the hole is made so as to direct the stream to have a tangential component.

Equipment
A large nail, an aluminum soda can, and some string.

Discussion
Using the nail, make a hole in the side of the can near the bottom. After inserting the nail, deflect it to one side, so that the hole will direct water flowing out of the can in that direction. Suspend the soda can by tying a string to the pull tab as close as possible to the center of the can, so that it is free to rotate about its axis when hanging from the string. Fill the soda can with water, suspend it by the string, and observe that after you uncover the hole allowing the water to flow, the can begins to turn more and more rapidly in the direction opposite the direction the water stream is deflected, thereby illustrating the principle of conservation of angular momentum—the same idea illustrated by a rotating lawn sprinkler.

This demonstration can also be used to make quantitative predictions regarding the rotating can. In particular, we can predict the angular acceleration of the rotating can, and the time for the initially stationary can to make one revolution. As will be shown later, the predicted time for one revolution if the can starts from rest is

$$t = \sqrt{\frac{4Mr}{\rho g h d^2 \sin \theta}},\tag{4.1}$$

where M is the mass of the water-filled can, r is its radius, ρg is the weight density of water ($\rho g = 9800 N/m^3 = 980 d/cm^3$), h is the distance the water level in the can is above the hole, d is the diameter of the hole, and θ is the angle the water stream makes with a normal to the side of the can.

The most difficult quantity for you to measure will probably be the angle θ. The best way to find it is to hold the can in your hand, with a protractor held with its edge against the can just above the water stream, and then look directly down through the protractor at the water stream. Notice that equation 4.1 not surprisingly predicts that when $\theta = 0$, the time for one revolution is infinity, because the can does not turn in this case. So, if you find that the can turns too rapidly to get a good measurement of the time for one revolution, reinsert the nail in the hole and reduce the angle θ.

However, be aware that if you reduce the angle θ too much, the torque due to twists in the string become relatively more important. Therefore, in order to get a reliable measurement when θ is small, you would need to release the can only after it hangs freely without turning due to string twists. You could accomplish this by covering the hole with tape to prevent flow, and letting the freely hanging can rotate to find its equilibrium orientation with the string fully untwisted.

The remainder of this section is devoted to deriving equation 4.1. We start by observing that the water exiting the hole at a speed v exerts a reaction force on the can given by $F = v dm/dt$, where the exit velocity v is given by $v = \sqrt{2gh}$, and the mass flow rate is given by $dm/dt = \rho v A$, where A is the area of the hole. Thus, the reaction force can be written as $F = \rho v^2 A$. This force exerts a torque on the can $\tau = Fr \sin \theta = \rho v^2 r A \sin \theta$, and results in an angular acceleration $\alpha = \tau/I$. Substituting the moment of inertia $I = Mr^2/2$ in the preceding equation yields $\alpha = 2\rho v^2 A \sin\theta/(Mr)$. Finally, if the can starts from rest, and makes a rotation through one turn (an angle of $\phi = 2\pi$ radians), the time required is $t = \sqrt{4\pi/\alpha}$. After substituting the previous result for α, and $\pi d^2/4$ for the area A, we find that the time t is given by equation 4.1.

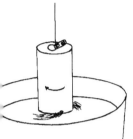

4.8 Inverse lawn sprinklers or *anti*-Hero engines

Demonstration
A simple version of the inverse lawn sprinkler (or *anti*-Hero engine) is found to rotate in the opposite direction to the normal lawn sprinkler.

Equipment
A one-inch (2.54 cm)-diameter steel ball, an aluminum soda can, a large nail to make holes in the can, some string, waterproof duct tape, and a bucket.

Discussion
In his delightful book, *"Surely You're Joking Mr. Feynman,"* Richard Feynman discusses an elementary physics problem involving the inverse of a standard lawn sprinkler.[4] An inverse sprinkler is simply one in which water is sucked in (rather than spewed out), by a submerged sprinkler that is free to rotate. Feynman notes that supposedly sound arguments can be advanced for each of the three possible outcomes: rotation in the same sense as the normal lawn sprinkler, rotation in the opposite sense, and no rotation. He also relates his comical ill-fated attempt to settle the matter experimentally. Since Feynman's book appeared, a number of papers have been published—see the list of references, for example, in the paper by Michael Collier, Richard Berg, and Richard Ferrell,[5] and additional references in an earlier 1989 paper by the first two authors.[6]

A number of the papers cited in reference[5] mistakenly claimed to find evidence for no rotation for the inverse sprinkler—as did this author in *Turning the World Inside Out.*[7] Nevertheless, Berg and Collier conclusively show that the inverse sprinkler does, in fact, rotate opposite to the direction of the normal sprinkler.[6] They attribute the mistaken negative results reported by others to friction, and even suggest making a version of their apparatus using extremely simple equipment.

To make the inverse sprinkler, use the same soda can from the previous demonstration on Hero's engine, and add a second hole in the side of the can opposite the first, using the large nail. Deflect the nail sideways, so that the hole causes the deflection of the second water stream to contribute to the same sign torque as the first one. Before submerging the can into a water-filled bucket, it is important to weight the bottom of the can by taping the steel ball to the hollow in the bottom of the can, otherwise the can will not enter the water in an upright position. Allow the empty can to slowly enter the water in the bucket, while hanging the can by the string tied to the flip top opener at a point close to the center. Keep the can away from the sides of the bucket, and you should observe that it rotates as the water enters the holes, and the can begins to sink.

The simplest explanation of the can's behavior is based on conservation of angular momentum. Assume that initially nothing is rotating. As the can enters the water, the flow streaming into the asymmetrically oriented holes causes the water in the bucket outside the can to develop some angular momentum as it swirls around the can and into the holes. The can and its contents must therefore develop an equal angular momentum in the opposite direction so as to keep the net angular momentum of the bucket and its contents zero.

4.9 Spinning a penny

Demonstration
If you spin a new penny on a very smooth hard surface, it will land with tails facing up at least 80 percent of the time.

Equipment
A fairly new penny and a smooth hard surface on which to spin it—such as a mica, glass, or plastic table top (glass is best).

Discussion
Due to its slight weight imbalance favoring its heads side, a penny experiences a slight gravitational torque as it spins. For very long spins that torque is nearly always decisive in leading the penny to land with its tails side up. To do the demonstration just keep track of how many spins lead to each outcome. It might be useful to keep track of the duration each spin lasts, so you can observe the correlation between spin duration and the probability of landing tails.

The extent of bias in spinning a penny may seem surprising in light of the lack of obvious bias for a tossed penny. The difference between the two cases arises because of the many randomizing factors operating for the tossed penny, such as your inability to accurately control the initial launch height, velocity, angle, and spin rate, which are more important in determining the outcome than the slight weight imbalance.

4.10 A fan of angular momentum conservation

Demonstration
A very light portable fan placed on its end on the OHP with its blade rotating in a horizontal plane will be seen to rotate opposite to the direction of its blade.

Equipment
A small battery-powered fan. Such fans can be obtained from scientific equipment companies, toy stores, and hobby shops.

Discussion
This demo works because the fan is sufficiently light so that the reaction torque to the torque on the rotating blade is sufficient to overcome friction and cause the fan to rotate in the opposite direction. Such reaction torques are especially important for helicopters which would spin opposite to the main rotor blade were it not for the stabilizing effect of the tail rotor. It is difficult to be quantitative about this demo unless we are able to estimate the moments of inertia

of the fan blade, I_1, and the rest of the fan, I_2. Presumably, we should find that, ignoring friction and angular momentum given to the air, the two rotation speeds of the blade and the rest of the fan are related by $I_1\omega_1 = I_2\omega_2$. This demo was shown at the Wisconsin South West Area Physics Sharing Group, and was related to me by Randy Elde.

4.11 The matchbook and the keys

Demonstration
A light and heavy mass tied to the opposite ends of a one-meter-long string passed over a pencil exhibits a most surprising behavior.

Equipment
A light and heavy mass (such as a matchbook and keys) tied to opposite ends of a one-meter-long string, and a pen or pencil to drape the string over.

Discussion
This demo may be considered a kind of Atwood's machine in which two masses hang on opposite sides of a pulley, except the pulley is replaced by a pen. Normally, in an Atwood's machine the light and heavy masses simply hang vertically. Here, however, the heavy mass (keys) should hang straight down a few centimeters below the pencil, but the matchbook should be held so that its string is nearly horizontal, with the matchbook held slightly below the level of the pencil. When you release the matchbook what do you predict will happen—won't the heavier keys just hit the floor?

The actual result is so surprising that you may want to make a prediction, and actually try the experiment before reading further. This demonstration is based on a stunt first suggested by Stewart James in the 1926 issue of a magician's journal, Linking Rings. Recent articles analyzing the demonstration from a physics perspective have appeared in the American Journal of Physics [8, 9, 10].

What you will find upon doing the experiment is that the keys, being by far the heavier mass, initially begin to fall, but that surprisingly they do not reach the floor. Instead, the string attached to the lighter mass winds itself around the pencil many times, so

Orbital Motion and Angular Momentum

that the resulting friction is enough to stop the descent of the keys. According to the AJP paper, this behavior is remarkably stable over a wide range of mass ratios from 10 to 200.[8]

The qualitative explanation of the behavior of the device is that as the heavy mass descends, and pulls the light mass toward the pencil, the rotational velocity of the light mass increases rapidly, first because it is swinging like a pendulum, and secondly because for a given value of its angular momentum, any decrease in radial distance must result in an increase in the angular speed—just remember the angular speedup of the figure skater when she brings her arms in.

Were it only for the pendulum motion, the light mass would swing only up to its original height. But, because of the second cause of increase in angular speed, the string goes over the top of the pencil, and wraps itself around a number of times—enough in fact to create an exceptionally large frictional force. (As shown in demonstration 3.10, the frictional force is an *exponential* function of the number of turns a string is wound around a cylinder.)

As far as a quantitative solution of the problem is concerned, although one can easily write down the equations obeyed by the two masses using Newton's second law—see, for example, the Sears article[10]—the solution of these equations cannot be given in closed form. Instead, one must choose some initial conditions and numerically integrate the equations on a computer. If you are particularly brave, you might want to do the demonstration using a delicate object for the heavier mass. I use a wine glass—though it might be a nice touch to literally use an (old and not too heavy) computer for the demonstration! In any case, if you do use a delicate object, be prepared for the possibility of something going wrong (remember Murphy's Law). Several times when I did the demonstration using a wine glass for the heavier mass, the string broke and the glass shattered on the floor.

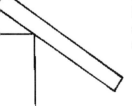

4.12 Jelly-side down

Demonstration
A piece of toast gently shoved off a table always lands jelly-side down.

Equipment
A piece of toast and some jelly.

Discussion
As shown by Ron Edge[11] and Darryl Steinert,[12] it is due to the laws of Newton not Murphy that a piece of toast gently knocked off a table always lands jelly-side down. As these authors independently show, the toast leaves the table with an angular velocity ω that depends only on the angle θ the toast has as it begins to slide off—typically about 30 degrees, as you can easily verify by putting the toast on an incline and varying its slope.

The analysis is somewhat messy, because as the toast slides off the table, both its moment of inertia and the torque acting on it due to gravity vary with time. The result of Edge's derivation (which has a small error in the final equation) is that the toast's angular velocity on leaving the table can be expressed as

$$\omega = 0.956\sqrt{\frac{g}{\ell}}$$

where ℓ is the length of the toast, and where it is assumed that the toast starts to slip at 30 degrees. The preceding equation yields $\omega = 10.0$ rad/s = 1.59 rev/s, for a 3.5-inch (8.9 cm)-wide piece of toast. If the toast slides off a horizontal surface having a height h, its time in the air is therefore $t = \sqrt{2h/g}$, and it makes ωt revolutions before impact. Adding the initial angle, 30 degrees (.083 rev), gives a total number of revolutions rev $= 1.59\sqrt{2h/g} + .083$. Thus, from this formula we can predict the number of revolutions that the toast should make for any given drop height.

Column one of the table shows four drop heights (in meters), and column four shows the predicted number of revolutions using the preceding formula. The second and third columns of the table, labeled "up" and "down," show the number of times the toast landed cleanly—either "jelly-side" up or down. The fourth column labeled "flip," shows the number of landings in which the toast did flips on landing—often the result of landing on or nearly on its edge. The data was taken by gently shoving a piece of

toast off a smooth horizontal surface at four different heights. To simulate the effects of jelly, some
aluminum foil was added mostly to one side of the
toast, so as to give it an asymmetry in mass. The foil
also helped keep the toast from disintegrating after many drops, and made its top and bottom easily
distinguishable.

h (m)	up	down	flip	rev
0.470	0	20	1	.57
0.775	14	4	33	.72
0.940	9	8	18	.78
1.105	20	0	6	.84

The data for each height agrees very well with the
prediction. For the lowest height (first row of table), the toast nearly always landed jelly-side down,
and the predicted rotation was just over half a revolution. For the middle two heights many edge-on
landings and flips occurred, which is just what would
be expected for a predicted rotation close to 3/4 of
a revolution. Apparently, for the greatest height, the
predicted rotation angle was sufficiently beyond 3/4
of a revolution for most landings to be jelly-side up.
One interesting facet of the data is that most toast
landings at a typical table or counter height (the middle two rows), resulted in landings involving flips.
The occurrence of these flips easily explains why the
toast usually lands jelly-side down, because given
a jelly-covered piece of toast, a flip is much more
likely to occur if the toast lands on the side without
the jelly—a plausible hypothesis you should feel free
to test.

The analysis described here applies only to the case
of a piece of toast that is gently shoved off the table
with near zero velocity—the way such accidents usually happen in real life. A piece of toast forcefully
shoved off the table will acquire a rotational velocity
that depends on its initial horizontal speed. Clearly,
for very high speed its rotational velocity would be
near zero, because the torque produced by gravity
has very little time to change the toast's angular momentum as it slides off the table—and hence it would
usually land jelly-side up.

4.13 String unwinding from a pole

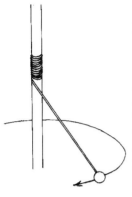

Demonstration
A weight hanging from a cord wound around a pole
unwinds in a predictable way as the weight rotates
around the pole in circles of increasing radius.

Equipment
A pole (such as a broomstick), a thick cord, and a
weight. A cord thickness of perhaps 0.5 cm will be
thick enough to avoid the problem of accidentally
overlapping the wound cord, yet thin enough to allow many turns to be wound on the stick.

Discussion
Wind the cord around the pole (carefully, so that the
cord does not lie on top of itself). After you have
wound one layer of cord, hold the pole upright, and
let the hanging weight slowly unwind the cord, as it
makes ever larger circles as the cord unwinds. Record
the elapsed time for the cord to unwind by various
numbers of turns. Let us call the time for the cord to
unwind by N turns t_N. As we will show, t_N is predicted
to be

$$t_N = [(2\pi a N)^{3/2} - \ell_0^{3/2}]F(\theta) \tag{4.2}$$

where a is the radius of the cylinder, ℓ_0 is the initial
length of cord attached to the hanging weight, θ is
the constant angle the unwinding cord makes with the
pole, and $F(\theta)$ is a function of θ, which can therefore
be treated as a constant for any number of turns, N.
As the string unwinds from the pole, observe that θ in
fact is approximately constant. To see how well your
observations of the predicted time to unwind various numbers of turns, N, agrees with equation 4.2,
you could simply plot $N^{3/2}$ versus t_N, which should
be a straight line. If you find that too tedious, here
is a much simpler check. Suppose we ignore the distance ℓ_0, then according to equation 4.2, we should
find that t_N is proportional to $N^{3/2}$. This proportionality implies that if we time the string to unwind by N
turns and then $2N$ turns, the ratio of those two times
should be $2^{3/2} = 2.82$. When I recorded the times for
the string to unwind by 14 turns and 7 turns, taking
four measurements for each case (in order to get an

uncertainty spread), I found a ratio of 2.73 ± 0.10, in good agreement with theory.

The rest of this section is devoted to deriving equation 4.2. Ignoring air resistance, the two forces acting on the revolving mass include the tension in the string T (acting at an angle θ with the vertical), and the force of gravity mg. Based on Newton's second law, we have for the horizontal and vertical force components $T \sin \theta = ma = mv^2/r$, and $T \cos \theta = mg$. Dividing the former equation by the latter yields $\tan \theta = v^2/rg = \omega^2 r/g$.

As the string unwinds, the weight spirals outward with a radial velocity $v_r = dr/dt = \omega a \sin \theta$, where a is the pole's radius. Substitution in the preceding equation yields the result

$$\tan \theta = \frac{r}{ga^2 \sin^2 \theta} \left(\frac{dr}{dt} \right)^2,$$

which upon rearranging terms gives

$$dt = r^{1/2} dr (ga^2 \tan \theta \sin^2 \theta)^{-1/2}.$$

Integrating the right side of the preceding equation from r_1 to r_2 gives the result

$$t = \frac{2}{3} (r_2^{3/2} - r_1^{3/2})(ga^2 \tan \theta \sin^2 \theta)^{-1/2}.$$

Finally, we use the fact that the radius of the circle r is related to the length of string ℓ through $r = \ell \sin \theta$, and also $\ell = 2\pi aN$. Upon substitution in the preceding equation, we get the result quoted in equation 4.2, with $F(\theta) = \frac{2}{3}(ga^2 \sec \theta)^{-1/2}$.

Notes

1. Synge and Griffiths, "Principle of Mechanics," p. 342 (1959).

2. R. Ehrlich, *Turning the World Inside Out*, p. 13.

3. R. Ehrlich and J. Tuszynski, The American Journal of Physics, **63**, 351–58 (1995).

4. R. P. Feynman, as told to R. Leighton; ed. by E. Hutchings, *Surely You're Joking Mr. Feynman*, Toronto; New York: Bantam Books, 1989.

5. M. Collier, R. Berg, and R. Ferrell, The American Journal of Physics, **57**, 349-55 (1991).

6. R. E. Berg and M. R. Collier, The American Journal of Physics, **57**, 654–57 (1989).

7. R. Ehrlich, *Turning the World Inside Out*, p. 34.

8. A. R. Marlow, The American Journal of Physics, **59**, 951–52 (1991).

9. D. J. Griffiths and T. A. Abbott, The American Journal of Physics, **60**, 951–53 (1992).

10. R.E.J. Sears, The American Journal of Physics, **63**, 854–55 (1995).

11. R. D. Edge, The Physics Teacher, September 1988, 192–93.

12. D. Steinert, The Physics Teacher, **34**, May 1996, 288–89.

Chapter 5

Conservation of Momentum and Energy

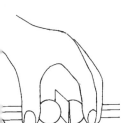

5.1 Momentum conservation on a ruler

Demonstration
Two balls, initially at rest, which push off against each other must recoil with speeds having the inverse ratio of their masses.

Equipment
A grooved plastic ruler, an index card, and three smooth metal or glass balls having a diameter of one inch (2.54 cm), which have masses approximately in the ratio of 1 to 3 to 3. For example, you could use two stainless steel balls and one aluminum ball. If you use a glass ball (marble) as one of the three, be sure that it is sufficiently round and rolls smoothly in the groove of the ruler.

Discussion
Place two of the balls in contact in the groove of the ruler, which should be placed on a horizontal surface such as an overhead projector. In general, overhead projectors are not exactly level, but there will always be some orientation of the ruler on the OHP so that a ball placed in its groove will not roll. Now place a folded index card sandwiched between the balls to serve as a low-force constant spring, and squeeze the card closed by finger pressure on the balls.

The advantage of using an index card to push the balls apart instead of a spring is that the card exerts a gentle force over a large distance, and is therefore less sensitive to nonsimultaneous finger releases— just be sure that you don't have sticky fingers! When you suddenly remove your fingers from the balls, the unfolding card will gently drive the balls apart with equal and opposite momenta. Therefore, if one ball is x times more massive than the other, its recoil speed should be the fraction $1/x$ of the lighter ball's speed.

A simple way to verify the preceding prediction is to initially place the balls so that their contact point is located at a point on the ruler where the lighter ball has x times the distance to travel to reach its end of the ruler than the heavier ball. In this case, after the balls recoil, they should reach their respective ends of the ruler at precisely the same instant. For example, the initial contact point should be at the midpoint if you use equal mass balls, and at the 3 inch (7.6 cm) mark if you use balls having a 3:1 mass ratio.

An alternative way to do the demonstration would be to treat the ball masses as unknowns, and find their mass ratio by varying the initial release point until you locate one from which the balls reach the ends of the ruler simultaneously. You could also find the uncertainty in the mass ratio by observing the range of release points that give that same result. If you then weighed the balls on a scale, you could see if the ratio of the ball's weights agrees with their mass ratio, given its uncertainty.

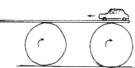

5.2 Walking the boat

Demonstration
As a spring-wound toy car placed on a plastic ruler propels itself forward, the ruler which rests on a pair of empty aluminum soda cans or stick pens moves backward. (The stick pens work better than the cans.)

Equipment
Two grooved plastic rulers, two stick pens, a one-inch (2.54 cm)-diameter stainless steel ball, and a small spring-wound toy car. The axle width of the toy car should be such that when the rulers are placed side by side, the car wheels can ride in the grooves of the two rulers. The best type of toy car is one which winds up when you pull it backward, so that you can easily control the approximate distance it will roll forward when released.

Discussion
A problem in many elementary physics texts concerns a person walking in a flat-bottom boat. As the person moves forward from one end of the boat to the other, the boat must move backward by an amount that depends on the ratio of the masses of person and boat. Assuming the lack of an outside net force, the momentum of the system must remain constant,

and therefore the location of the center of mass does not change during the motion.

Rather than using a floating platform to do the demonstration, Akio Saitoh has suggested an alternative version using a rectangular piece of lucite resting on a bunch of BB's.[1]

However, the version suggested here dispenses with the lucite slab, and uses rulers resting on stick pens instead. To do the demo, place the two grooved plastic rulers side by side on the pens which should be free to roll on a horizontal surface, such as an overhead projector. The toy car should be placed on the rulers so that its wheels sit in the ruler grooves. The rulers should be placed on the pens in such a way that they can roll the greatest distance possible with the rulers resting on them. (Note that you may find that the pens keep rolling after the car has stopped, in which case you probably have not placed them on a level flat surface.)

To do the demonstration you will want to observe how far the car moves forward on the ruler x_c, and also how far the ruler itself moves backward, x_r. Of course, relative to the ground, the car moves only a distance $x_c - x_r$. One other slightly tricky point concerns the motion of the pens on which the ruler rests—they roll backward only *half* the distance of the ruler (as discussed in demonstration 4.2). If we call the masses of the car, ruler, and pen m_c, m_r, and m_s, respectively, the requirement that the center of mass not move can therefore be expressed as:

$$m_c(x_c - x_r) = m_r x_r + m_s \frac{x_r}{2}$$

which can be solved for x_r to yield:

$$x_r = \frac{m_c x_c}{m_r + m_c + \frac{m_s}{2}}$$

—a result you could try to confirm if you measure x_r, x_c, and the three masses. Quantitative agreement is probably unlikely, given any surface irregularities or slope.

Alternative version
Originally, I tried an alternative simpler version of this demo. Instead of using a toy car, we can use a one-inch (2.54 cm)-diameter steel ball which is rolled

in the groove of one of the rulers while the ruler rests on the two pens. Release the ball from a point closer to one of the pens than the other, and it will roll back and forth about a midpoint, because the ruler flexes under the weight of the ball, and the ball finds itself in a potential well. Given that the steel ball is more massive than the toy car, you might expect to observe an appreciable recoil motion of the ruler and pens as the ball moves back and forth. But, interestingly, there is no motion whatsoever.

The outside force responsible for the ball's oscillation is gravity. But, since gravity has no horizontal component it cannot affect the horizontal location of the center of mass. You might therefore expect that as the ball oscillated the ruler should oscillate opposite to it. Unfortunately, rolling function for the ball is appreciably less than for the pens, so the ruler remains stationary.

5.3 Colliding coins and transverse momentum

Demonstration
By projecting one coin into a second stationary coin, you can make a quantitative test of the law of conservation of momentum.

Equipment
Two coins (two quarters or a quarter and a nickel are particularly suitable), and a ruler.

Discussion
Everyone has probably flicked one coin into another on a table top, and observed the kinds of collisions that result—from head-on, to glancing, to something in between. Amazingly, this demonstration can be made into a *quantitative* check on the law of conservation of momentum—or at least one component of momentum.

During a collision between two coins when one is projected by hand into another, momentum is conserved both along the direction of motion of the projectile coin (the x-axis), as well as the transverse direction (the y-axis). Here we shall describe a method to verify that the transverse component of momentum is conserved. Unfortunately the method cannot check on the conservation of the x-component, because we cannot easily manually control the speed with which the projectile coin is launched. The method sounds more complicated than it actually is. But, essentially, you want to be able to measure four quantities: the distances each coin travels after their impact, and the angles they make with respect to the projectile coin.

Start by placing the projectile coin near the bottom edge of a piece of paper (defined as position 1), and the target coin near the center of the paper (position 2)—see figure. With a pencil in contact with the edge of each coin, locate the positions precisely by drawing a circle around each one. (If you are doing the demonstration on an overhead projector, you would use a transparency blank instead of paper.) With one finger, flick the projectile coin toward the target with enough speed so that the two coins after their collision each travel far enough on the paper to make a reliable measurement of their path length after collision. This requirement means that you wish to avoid collisions that are either extremely glancing or nearly head-on, because such types of collisions result in one of the coins traveling only a very short distance after collision. Circle the position where the target coin (position 3), and the projectile coin (position 4) come to rest after collision.

Before you can make measurements of the distance each coin traveled after the collision, you need to first find the location of position 5 where the projectile coin was at the instant it made contact with the stationary target coin. Here are two possible strategies for locating position 5: (*a*) put 5 next to 2 and in a straight line joining positions 2 and 3, and (*b*) put 5 next to 2, so that the line joining 5 and 2 is at right angles to that joining 5 and 4 (see figure).

Had the collision between coins been elastic, and the coins of equal masses, these two rules would be equivalent, because in that case, the coins after collision would travel along paths 90 degrees apart. But, since the collision is not elastic, and the masses may not be equal, we shall simply assume that we can locate position 5 by taking the average of the positions suggested by the two preceding rules—that is, just locate position 5 midway between what you get by applying the two rules.

Once you have circled position 5, draw a line connecting the centers of positions 1 and 5, which defines the x-axis or the direction of the projectile. Also draw a line from the center of position 2 that is parallel to the x-axis (line x'.) Finally, draw lines connecting the centers of positions 2 and 3, and 5 and 4. The length of these two lines represent the distances s_1 and s_2 traveled by the target and projectile coins after the collision. The velocities of the two coins immediately after the collision can be expressed as $v_1 = \sqrt{2as_1}$, and $v_2 = \sqrt{2as_2}$, where we have assumed that the same acceleration brings each coin to rest, which is true if the coefficient of friction is the same for both coins.

Dividing the two previous equations yields $v_1/v_2 = \sqrt{s_1/s_2}$. Since we don't really need the absolute velocities, but only their ratio, let us choose an arbitrary system of units such that $v_1 = s_1$ in magnitude. The previous equation would then imply that in this system of units $v_2 = \sqrt{s_1 s_2}$. Thus, based on the two measured path lengths we have a way of finding the two velocities, which you should draw as vectors along the two coin paths—see figure. But, since we are testing for conservation of *transverse* momentum, we need to find the y-component of each velocity.

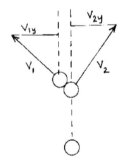

The transverse component is found by directly measuring the length of the y-component of each vector. To test for transverse momentum conservation, you need to see if the values of v_{1y} and v_{2y} equal the inverse ratio of the coin's masses. (Based on the figure shown, what would you judge the mass ratio to be?) If you find a departure from momentum conservation in a given case, make measurements on a number of collisions to see how much variation you get using the same set of coins—then try a pair of coins having another mass ratio. My measurements yielded consistency with momentum conservation to within a few percent! That is, when using equal mass coins, the two transverse velocity components were usually within a few percent of each other, on the average.

**Conservation of
Momentum and
Energy**

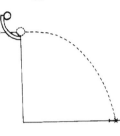

5.4 Projectile trajectory on an incline

Demonstration

If you roll a ball with a fixed initial velocity on an inclined plane, it follows a predictable parabolic trajectory that can be computed using conservation of energy.

Equipment

A small ball, such as a steel ball one inch (2.54 cm) in diameter, but don't use a marble, because they are usually too irregular. For the inclined plane, use either a note pad or a piece of transparency if you want to do the demonstration on the OHP. The transparency will be placed on the tilted OHP to create an inclined plane. In either case, you will also need a curved piece of track, such as a plastic ring of the type that formerly was inserted in magnetic tapes to allow mainframe computers to write on them. Most computer centers still have such rings. A possible substitute for such a ring would be an embroidery hoop. Cut the ring into quarters, so as to produce 90-degree arcs.

Discussion

The demonstration of the parabolic trajectory of a ball rolling on an inclined plane has a long history.[2] However, in the usual form of this demonstration, it is not possible to predict the *specific* parabola a ball will follow without knowledge of

- the angle of the incline
- the horizontal velocity with which the ball is launched
- the mass distribution of the ball, which determines the formula to be used for its moment of inertia, and hence its acceleration down the incline.

The present version of this demo allows the parabola to be predicted without knowledge of any of these quantities. To construct the demo, draw a picture of a vertical cliff on the note pad (or a transparency blank), with the vertical drop accurately oriented along the long dimension of the note pad. The base of the cliff should be drawn straight across the pad about a centimeter from the bottom of the paper. The edge of the cliff should be positioned about 5 centimeters from the top of the pad and 5 centimeters from the left side, so as to leave enough room in the upper left corner of the pad to place the 90-degree arc of track cut from the plastic ring there. As shown in the figure, the arc should be placed so that one end is right at the edge of the top of the cliff, and oriented so that a tangent to that end is horizontal. Now, tilt the note pad so that its top end is propped up by some books. Be sure that a line running across the note pad remains horizontal, and be sure you are not using a note pad that flexes under its own weight when propped up at an angle.

If you hold the arc of track in place at the top of the cliff with one hand, you can place a ball at the top of the track with the other hand, and let it go. If the track is oriented properly, and your finger holding the track in place does not get in the ball's way, it will roll along the arc of the track, and leave the edge of the cliff with a velocity that only has a horizontal component. The value of that initial velocity can be found using conservation of energy as $v_0 = \sqrt{kgh}$, where h is the height of the top of the arc of track above the edge of the cliff, and $k = 10/7$ for a solid sphere. (For a frictionless mass sliding down the track, without rotational energy, we would have used $k = 2$.)

Once the ball leaves the edge of the cliff it follows a parabolic arc, and it falls down the incline with an acceleration $a = \frac{kg}{2}$. The time for the ball to descend a vertical distance y to the base of the cliff is therefore $t = \sqrt{2y/a} = \sqrt{4y/kg}$. In that time, it should travel a horizontal distance $x = v_0 t = \sqrt{kgh}\sqrt{4y/kg} = \sqrt{4hy}$. Note that the predicted range is independent of g, which is important here, because the effective value of g on an incline depends on the angle of the incline, so our result does not depend on that angle.

A nice way to do the demonstration would be to compute the predicted projectile range from the preceding formula, and make an X located a distance from the base of the cliff where the ball is predicted to land—or better yet, draw the entire dotted parabola shown in the figure, which is calculated from $x = \sqrt{4hy}$.[3] You could then observe the actual path of the ball if it is dipped in ink or some other colored fluid just prior to being rolled. As noted earlier, in contrast to the usual demo of a parabolic trajectory on an incline, in this demo the

**Conservation of
Momentum and
Energy**

specific parabola does not depend on the angle of the incline.

5.5 Ballpoint pen test of energy conservation

Demonstration
A retractable ballpoint pen can be used to test the law of conservation of energy by launching the top part of the pen upward by the compressed spring and comparing its measured maximum height with the theoretical prediction.

Equipment
A retractable ballpoint pen and a meter stick. You need to use the type of ballpoint pen that can be disassembled into three main pieces: the ink-carrying stem, and front and back plastic pieces that screw into each other.

Discussion
Here is the proper technique for launching the front part of the pen upward. Disassemble the pen, but leave the spring on top of the ink-carrying stem, which should be held vertically with its bottom end on your desk. Place the front plastic part of the pen on top of the spring, and compress the spring by grasping the lower end of the plastic part between your thumb and index finger, and pulling it down as far as it will go. Finally, if you suddenly release your finger grip on the bottom of the plastic part, it will jump up in the air as the spring expands.

You can easily estimate the maximum height of the vertically launched plastic part if you have placed a vertical meter stick next to it. If you keep your head at the anticipated maximum height, and launch the projectile so that its distance from your face is the same as the meter stick—so as to avoid parallax—you can probably estimate the height to the nearest centimeter. Obviously, if you make a series of repeated measurements, you need to accept only cases where the launch is essentially vertical, and the projectile does not strike your hand holding the meter stick.

To predict the maximum height, h, that the top of the projectile rises in the absence of friction, we need only equate the elastic potential energy in the fully compressed spring to the potential energy of the plastic part at the top of its path, assuming it rises straight up, i.e., $mgh = \frac{1}{2}kx^2$, so that $h = \frac{kx^2}{2mg}$. To obtain a value for k needed in this equation, we must see how much force is required to fully compress the spring. This force measurement can be easily accomplished by placing the bottom of the stem against one pan of a double pan balance and pressing the plastic part down until the spring is just maximally compressed.

If we let M represent the amount of mass you need to add to the other pan to achieve balance, then the force constant is given by $k = Mg/x$, and the predicted maximum height becomes $h = \frac{Mx}{2m}$. I found values $m = 2.8\,g$, $M = 250\,g$, and $x = 1.7\,cm$, yielding $h = 76\,cm$. The greatest uncertainty in the computed height probably comes from the uncertainty in the distance x. If we assume that x can be reliably measured to no better than half a millimeter, and we ignore the uncertainties in the masses, we find for the predicted maximum height $h = 76 \pm 2\,cm$.

The results from measurements of the maximum height initially varied dramatically depending on the particular pen used, and on exactly how tightly the bottom of the spring was jammed onto the stem. Apparently, a great deal of mechanical energy is lost to friction if the spring is able to slide past the bulge on the stem at the place where the stem is crimped. It is instructive to deliberately push the bottom of the spring past the bulge and observe how dramatic a reduction in height results. The key to achieving good results showing little energy loss is to crimp the stem a bit more at the point of the bulge, so that the spring cannot slide onto or past the bulge.

With this slight modification of the pen, the mean of sixteen launches gave a maximum height of $73.8 \pm 2.0\,cm$, in excellent agreement with the predicted height. One could obviously extend this experiment to include projectiles of various masses, and see whether the results continue to agree. Heavier projectiles can easily be made by adding pieces of clay to the plastic piece. Lighter projectiles can be made by sawing a portion of the plastic piece off. Presumably, one should find that for very light projectiles, the maximum height predicted from the above formula is in error, because it ignores the kinetic energy of the spring at the instant the plastic piece leaves the spring. (This demo was described

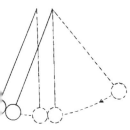

in an article appearing in The American Journal of Physics.)[4]

5.6 Inelastic collisions using "Newton's Cradle"

Demonstration
The widely familiar "executive toy" consisting of a row of swinging balls can be used to test momentum conservation in inelastic collisions.

Equipment
A "Newton's Cradle" toy, a small piece of clay, and a transparency made from a piece of graph paper if you want to do the demonstration on the overhead projector.

Discussion
The toy popularly known as "Newton's Cradle" consists of a row of five balls hanging by strings from a frame. The fascination of the toy comes from its simple dynamics: when one ball is pulled aside and released, its impact with the row of balls causes the one at the other end to fly off, leading to a long-lasting cycle of single balls flying off each end, with the other balls remaining largely motionless. It is clear that the low damping of this cycle depends on the high coefficient of restitution of the balls, meaning that the collisions are nearly elastic—an issue considered in the next demonstration. Here, we shall consider how the toy may be used to illustrate momentum conservation in completely *in*elastic collisions.

In order to make quantitative observations, place a transparency made from a piece of graph paper on the OHP, and place the toy on top, with one axis of the graph paper along the direction the balls swing. Pull all but two of the balls aside, and rest them on an improvised shelf made from books, so that you can examine the collisions between a single pair of balls. Put a tiny amount of clay on one of the two balls, so that they will stick together on impact. Now, pull one ball aside and release it from a specific distance x_1

from the other stationary ball as seen on the transparency graph paper. After the balls collide and stick together, observe how far x_2 the two balls move after the collision before they swing back. If you cannot get a good reading of the maximum displacement on the first swing, see what you find on the second or third swing, as the two balls swing together. Be sure that you measure x_1 and x_2 as the distance a particular point on the balls moves.

Since the collision between the two equal mass balls is completely inelastic, momentum conservation requires that immediately after collision, their common velocity be half the initial velocity of the first ball just before collision. If we apply energy conservation during the two balls' swing following the collision, $mgy = \frac{1}{2}mv^2$, we find that the maximum height they can rise to is proportional to the square of their velocity—so the two balls together should rise only a quarter as high as the first ball was initially.

But, remember we don't directly observe the ball's heights on the OHP—only their projected (horizontal) displacement, x. For small angle swings, we may assume that the height y is proportional to x^2—in other words, a circular arc is approximately a parabola. Therefore, if the velocity is halved after the inelastic collision, and the height is quartered, it would be predicted that x should be halved, so that we predict $x_1/x_2 = 2$. See if you begin to observe any departures from this prediction, as the initial angle of swing is increased.

Another way to do the demonstration is in the CM reference frame of the colliding balls, for which the prediction is zero velocity following the inelastic collision. It is fairly easy to obtain a collision in which the balls approach each other with equal and opposite velocities using a 5-centimeter-long rolled up piece of index card. Place the length of rolled-up piece of card between the two balls. You should be able to gently place the card and have it held in place between the balls as their weight presses slightly against its ends—see figure for next demo. When the balls are motionless, remove the card without disturbing them, and observe that they symmetrically swing toward one another, and become stationary on impact.

5.7 Coefficient of restitution

Demonstration
The widely familiar "executive toy" consisting of a row of swinging balls can be used to measure the coefficient of restitution during collisions.

Equipment
A Newton's Cradle toy and a transparency made from a piece of graph paper if you want to do the demonstration on the overhead projector.

Discussion
As in the previous demonstration, let us consider the problem of a collision between only two balls—with the remaining balls of the toy resting on an improvised shelf. But, this time we wish to see how close to being elastic the collisions are, so we do not introduce any clay between the balls. In fact, if you added a lump of clay in the previous demonstration, be sure to wipe the balls clean, because even a surface film of clay can seriously reduce the measured elasticity of the collision.

Suppose we have an initial state in which both balls are pulled aside, and released from identical

Conservation of Momentum and Energy

angles, so that the lab and center of mass systems become identical. You can accomplish this situation using a rolled-up piece of index card placed between the balls. For this symmetric situation, each ball undergoes exactly one-half cycle of a full pendulum swing between impacts, so that the frequency of impacts (for small angle swings), is predicted to be $f = \frac{1}{\pi}\sqrt{g/\ell}$, where ℓ is the length of the strings.

A very convenient way you can confirm that the predicted frequency is correct is to activate a metronome at the predicted frequency, and observe that the ball collisions occur at a steady frequency that exactly match that of the metronome. By placing the apparatus on an overhead projector with a ruled transparency beneath it, one can observe how the amplitude of the maximum ball separation decays with time.

The coefficient of restitution, ϵ, may be defined as the ratio of the ball speeds after collision to those before collision in the center of mass system, so that $0 \leq \epsilon \leq 1$, where the limiting values correspond to completely inelastic and elastic collisions, respectively. To reliably measure the coefficient of restitution here we need to observe the result of many successive collisions, because so little energy is lost during one collision. If one considers the result of N collisions of a pair of equal mass balls, the result is to reduce the initial common speeds of the balls by the factor ϵ^N—assuming that a constant fraction of the mechanical energy is lost with each collision. If no mechanical energy were lost during each ball's swing, we could relate its velocity just after impact v to its maximum angle of swing θ. Using conservation of energy, we find that $\frac{1}{2}mv^2 = mg\Delta y$, where $\Delta y = \ell - \ell\cos\theta$, so that

$$v^2 = 2g\ell(1 - \cos\theta). \tag{5.1}$$

But, if we assume more realistically that some energy is lost during the swing as well as during the collision, equation 5.1 becomes instead

$$v^2\delta = 2g\ell(1 - \cos\theta) \tag{5.2}$$

where δ is the fraction of mechanical energy retained during a swing. Hence, based on equation 5.2, and the definitions of ϵ, and δ, we can express the quantity $\epsilon^2\delta$ in terms of the initial angle of each ball prior to release, θ_0, and their maximum angles after N collisions, θ_N.

$$\left(\epsilon^2\delta\right)^N = \frac{v_N^2}{v_0^2} = \frac{1 - \cos\theta_N}{1 - \cos\theta_0} \tag{5.3}$$

so that we have

$$\epsilon^2\delta = \left(\frac{1 - \cos\theta_N}{1 - \cos\theta_0}\right)^{1/N}. \tag{5.4}$$

Note that $1 - \epsilon^2$ represents the fractional mechanical energy loss during a collision, and $1 - \delta$ represents the fractional mechanical energy loss due to frictional effects between collisions. The quantity we actually observe as the pair of balls repeatedly collide is not θ, but rather the maximum horizontal displacement x of each ball during its swing, which is related to θ by

$$\theta = \sin^{-1}(x/\ell). \tag{5.5}$$

Experimental results
To find experimental values for ϵ and δ you need to first count the number of impacts N for the initial horizontal ball displacement to be reduced to half its initial value to find $\epsilon^2\delta$ using equations 5.4 and 5.5. Then, after that you can find δ separately, by seeing how many swings (without impacts) are needed for the ball displacement to be reduced in half.

Here are the results of one such experiment. I observed that $N = 158 \pm 10$ collisions were required for the initial horizontal ball displacement ($x_0 = 1.8\,\text{cm}$) to become reduced to half ($x_N = 0.9\,\text{cm}$). Actually, it was a bit too tedious to count up to 158 impacts, so I simply measured the time, and found N using the computed frequency of impacts, which was previously found to be reliable—based on the metronome measurement noted earlier. Using the measured displacements, we can use equation 5.5 to convert them to angles, and then use equation 5.4 to find $\epsilon^2\delta = 0.9912 \pm 0.0008$.

In order to determine ϵ and δ separately, we observe the effects of frictional damping on a freely swinging ball with no collisions ($\epsilon = 1$), as its horizontal amplitude decays the same amount as before (from $x_0 = 1.8$ cm to $x_N = 0.9$ cm). The number of half swings for this amplitude reduction to occur was found to be $N = 265 \pm 15$, for which equation 5.4 yields $\delta = 0.9974 \pm 0.0001$. Combining this value of δ with the previous value of $\epsilon^2\delta$ yields $\epsilon = 0.9969 \pm 0.0004$, meaning that only $100 - 99.69 = 0.31$ percent of the ball's mechanical energy is lost in each collision. But don't expect to find the same value when you do the experiment, because the coefficient of restitution can depend on the radius and initial velocity of the balls, as well as their composition. (This demo is based on an article I wrote for The Physics Teacher.)[5]

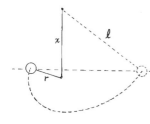

5.8 The interrupted pendulum

Demonstration
A swinging pendulum whose string is stopped by a fixed peg illustrates the principle of conservation of energy.

Equipment
A board (a little over a meter in length) from which a pendulum bob is hung from a long nail placed near the top of the board. Make the pendulum length $\ell = 1$ meter, and place a second long nail a distance $x = 0.7$ meter directly below the top nail.

Discussion
Press the board firmly against the floor, and allow the pendulum to swing so that the string is stopped by the bottom nail. First release the pendulum bob at an initial angle such that the height of the pendulum bob is at or below the height of the lower nail. You should observe that at the end of its swing it rises to the same height that it started at, even though the string is stopped by the nail. This outcome could be made clear if you draw a line across the board representing the starting height of the pendulum. The reason the bob rises up to its initial height is that the force of the nail on the string does no work (since there is no movement of the string there), and therefore the mechanical energy remains constant.

Let us predict the minimum initial height of release for the pendulum that would allow it to keep circling the bottom nail. In order for the pendulum just to complete the circle when it is at the top, the tension in the string T must satisfy $T > 0$ there. But, by Newton's second law, we have: $T + mg = mv^2/r$ at the top of the circle, so that its velocity at the top must satisfy $v > \sqrt{gr}$, and its kinetic energy there must satisfy $\frac{1}{2}mv^2 > \frac{1}{2}mgr$. Let us assume that the potential energy is zero at the top of the circle. At its initial height y above the top of the circle, the stationary pendulum therefore had a potential energy $mgy = \frac{1}{2}mv^2$. So combining this last result (from conservation of energy), with the previous inequality, we find $y > r/2$ for the minimum release height above the top of the circle—or a minimum height of 1.5 radii above the bottom nail. You should find that for lower release heights, the pendulum doesn't quite complete the circle.

For a final part of the demonstration you might put some clay around the nail. Now when the string hits the clay-covered nail during its swing, the force stopping the string *does* do work (since it acts through a nonzero distance), and some mechanical energy will be lost, with the result that it will no longer rise up to the same height as its initial level. More details on this demonstration can be found elsewhere[6, 7].

5.9 Dropping two rolls of toilet paper

Demonstration
If you drop two rolls of toilet paper while holding on to the end of one roll, they will hit the floor at the same instant if their initial heights have a specific ratio.

Equipment
Two rolls of toilet paper and a meter stick.

Discussion
Drop a roll of toilet paper from a height y_1 and it will hit the floor in a time $t_1 = \sqrt{2y_1/g}$. If you drop a second roll of toilet paper while holding on to the end of the roll, its downward acceleration a will obviously be less than g because of the upward force acting on the roll. If the second roll is dropped from a height y_2, its time of fall is $t_2 = \sqrt{2y_2/a}$. In order that the two rolls hit the floor at the same time ($t_1 = t_2$), we must

Conservation of Momentum and Energy

have that the ratio of the two initial heights satisfy $y_2/y_1 = a/g$. As we will show below, the acceleration a is given by $a = 2g/(3+R^2)$, where R is the ratio of the diameter of the hole in the roll to the diameter of the roll itself. Therefore, we may predict that to get the two rolls to hit the floor at the same instant, the ratio of their release heights must be $y_2/y_1 = 2/(3 + R^2)$, which you should be able to confirm experimentally. For example, for a full roll of toilet paper, $R = 0.38$, so that the predicted height ratio for simultaneous impacts is therefore 0.64.

In the remainder of this section we will show that the acceleration a has the value given above. First, it can be easily shown that the moment of inertia of a cylinder of mass M and radius r_2, which has a hole of radius r_1 is given by $I = \frac{1}{2}M(r_2{}^2 + r_1{}^2)$. (Notice that this formula gives the correct limiting cases for $r_1 = 0$ [a solid disk] and for $r_2 = r_1 + dr$ [a ring.]) The preceding formula gives the moment of inertia about an axis through the center of the cylinder. When you drop the toilet paper roll while holding on to one end, the roll is momentarily rotating about an axis at the edge of the roll, where its moment of inertia is given by $I = \frac{1}{2}M(r_2{}^2 + r_1{}^2) + Mr_2{}^2 = \frac{1}{2}M(3r_2{}^2 + r_1{}^2)$, based on the parallel axis theorem.

The angular acceleration of the roll during its fall can be found from $\alpha = \tau/I$, where the net torque is given by $\tau = Mgr_2$. Using the previous result for the moment of inertia, we have $\alpha = 2gr_2/(3r_2{}^2 + r_1{}^2)$. Finally, since the acceleration of the center of mass, a, is related to the angular acceleration of the roll by $a = \alpha r_2$, we get the result $a = 2gr_2{}^2/(3r_2{}^2 + r_1{}^2) = 2g/(3 + R^2)$, where $R = r_1/r_2$.

Notes

1. A. Saitoh, The Physics Teacher, May 1985, 316–17. One disadvantage of this method, however, is that the OHP needs to be leveled very accurately, because otherwise the lucite tends to wander once it starts moving—a problem which could be alleviated by resting the lucite slab on cylinders, such as pens or drinking straws, rather than spherical BB's.

2. See, for example, the discussion by Thomas Greenslade Jr., in his article in The Physics Teacher, **34**, 156–57 (1996).

Conservation of Momentum and Energy

3. Actually, it is unrealistic to expect the ball to follow the theoretical parabola exactly, so you may wish to draw two parabolas, thereby showing an uncertainty band. The main source of uncertainty is probably due to variations in the value of h, the distance above the top of the cliff from which you release the ball, and also errors in track orientation, which cause the ball to have some vertical velocity component when it leaves the edge of the cliff. If we just take the former source of uncertainty into account, for any given y, the uncertainty in x (the horizontal distance between the two parabolas at a given y), would be $\Delta x = x \Delta h / h$.

4. R. Ehrlich, The American Journal of Physics, **64**, 176 (1996).

5. R. Ehrlich, The Physics Teacher, **34**, 181–83 (1996).

6. H. Wood, The Physics Teacher, **32**, 422–23 (1994).

7. P. Robinson, *Conceptual Physics Lab Manual*, Reading, MA: Addison-Wesley, pp. 75–76 (1987).

Chapter 6

Fluids

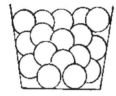

6.1 Volume is not conserved

Demonstration
A considerable amount of water can be added to a cup "filled" with marbles.

Equipment
A transparent plastic cup filled with marbles, a ruler, and a graduated cylinder.

Discussion
Measure the diameter of a marble, D, and the volume of the cup. Challenge your audience to figure out how many marbles the cup can contain based on these two quantities. A good way to measure the average diameter of a marble would be to line up a row of marbles against a ruler, and measure the length of some number of them. The simplest way of measuring the volume of the cup would be to measure how much water it contains when full using a graduated cylinder.

Now, suppose the marbles are added to an empty cup until they fill the cup, with none projecting above the rim. If there were no empty spaces between the marbles, the number of marbles you could fit in a cup, N_0, would be equal to the volume of the cup divided by the volume of one marble. Try the experiment, and see how many marbles you, in fact, can put in a cup. According to theory, a random packing of spheres should occupy 64 percent of the volume, so your observed number should be in the vicinity of $0.64N_0$. (The closest possible packing of spheres occupies 74 percent of the volume.)

One final part of the demonstration would be to see how much water could be added to the cup "filled" with marbles. You should, of course, find that this volume equals the volume of the cup minus the total volume of all the marbles it contains. But, in a more important sense, volume is not really "conserved," because you can add water to a cup "completely filled" with marbles. This nonconservation of volume occurs in many cases when you combine nonreacting substances whose molecular sizes significantly differ. For example, if you were to add equal volumes of water and alcohol, the mixture would have somewhat less than twice the volume—because the smaller molecules are filling some of the spaces between the larger ones. (For exactly the same reason, if you added half a cup of small beads to half a cup of marbles, the result would be less than a full cup of beads and marbles.)

6.2 Floating ice cubes

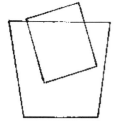

Demonstration
When ice cubes floating in a filled cup of water melt, the water does not overflow.

Equipment
Some ice cubes floating in a small water-filled plastic cup.

Discussion
If you place the filled cup on the OHP, it will be obvious to everyone if the cup overflows when the ice melts. However, the melting ice should *not* cause any overflow during melting, because as we shall see, the volume of water created equals the submerged volume of the ice, not its total volume.[1]

How can we prove that a melting ice cube results in no overflow? We start with Archimedes' principle, which says that the buoyant force on the ice cube equals the weight of the displaced water: $B = \rho_{water} V_s g$, where V_s is the volume of ice that is submerged. Next, since the net force on the ice is zero, we can equate the buoyant force to the weight of the ice: $\rho_{water} V_s g = \rho_{ice} V_{ice} g$. Finally, once the ice melts, the ratio of the resulting water volume to the original volume of ice is just the inverse ratio of the densities: $V_{water}/V_{ice} = \rho_{ice}/\rho_{water}$, which, using the preceding equation, must equal V_s/V_{ice}—showing that we must have $V_{water} = V_s$.

6.3 Buoyant force on your finger

Demonstration
If you dunk your index finger into a cup partially filled with water sitting on a scale, the scale reading will increase as you submerge your finger in reaction to the buoyant force on your finger.

Equipment
A digital scale and a cup of water.

Discussion
Put a partially filled glass of water on a scale. If you gently insert your finger into the water, will the scale reading increase? When you dunk your finger into the water, it experiences an upward buoyant force, given by $\rho_{water}Vg$, where V is the volume of your finger that is submerged. The reaction to this force—the force with which your finger presses down on the water—must show up as an increase in the scale reading. In fact, you could use this observation to find the submerged volume of your finger. It might be interesting to compare this measured value with what you obtain from a calculation of your finger's volume when approximating it by a cylinder.

6.4 Four sucking problems

Demonstration
You cannot suck a liquid through a straw under a number of circumstances.

Equipment
A number of straws, some tape, a cup of water, and a soda bottle having a cap in which you have made a hole just big enough for the straw to fit through.

Discussion
When you suck a liquid up the straw, it is actually being *pushed* up the straw by the pressure of air on the surface of the liquid. In fact, in order to force the liquid up the straw, the pressure on the liquid surface must exceed the pressure inside your mouth by an

amount that is at least as great as the pressure due to the weight of the liquid in the straw. This concept is illustrated in the following four situations.

Problem 1. Here is a way to literally measure how big a sucker you are. Tape a number of straws together end to end—being careful to avoid leaks—and see how many straws you could suck water up through. You will probably need to put a glass of water on the floor and stand on a chair if your straw is very long. Since one atmosphere is equivalent to a height of about 10 meters of water, for each meter you can suck it up, you are creating a pressure reduction inside your mouth equal to a tenth of an atmosphere. For straws longer than a certain length, the pressure of the water column filling the straw would exceed the maximum pressure reduction you can create inside your mouth, so you cannot suck it up.

Problem 2. Put two straws in your mouth, but let only one of them have its lower end submerged in a glass of water, with the other one in the air. No matter how hard you try to suck on the straws, you will not get any water to drink. You can only suck a liquid up a straw if you create a pressure inside your mouth less than atmospheric pressure. But such a reduction is impossible if one of the straws in your mouth has its other end in the air.

Problem 3. Put one of the straws into the hole in the cap of the soda bottle, which has some water in it. No matter how hard you try, you will not get much to drink, because as you reduce the pressure at the top of the straw by sucking, and the liquid starts to rise, the pressure inside the bottle decreases (as the volume of air expands), and finally it is no longer great enough to keep pushing the liquid up the straw.

Problem 4. A variation on problem 3 would be to use one of those small cardboard juice containers that comes with a tight fitting flexible straw. If you insert the straw, and suck continuously, you will initially get some juice, as atmospheric pressure squashes the cardboard container. But at some point you get no more juice, because a greater sucking pressure than you can exert is needed to crush the container more fully. Of course, if you stop sucking and let air in through the straw, you can then get another drink once the cardboard container returns to its original shape.

6.5 Egg in a water stream

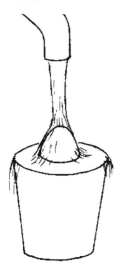

Demonstration
An egg under the stream of water from a faucet is actually pulled upward as the water flow rate increases.

Equipment
An egg floating in a glass of water.

Discussion
It is most surprising to observe an egg floating in a glass of water rise upward as the flow rate of water hitting it increases. This demonstration is described in Jearl Walker's *The Flying Circus of Physics*.[2] However, contrary to that description, which implies that the egg rises at some threshold flow rate, the rise appears to be a continuous function of flow rate.

What is the explanation of this strange behavior? One might logically assume the force of the descending water should drive the egg down further as the flow rate increased. The reason that the opposite occurs has to do with the extremely streamlined shape of an egg. As a result, the force of the falling water on the egg is far less than it would be if it had, say, a flat shape. The water is simply deflected aside by the egg, and its velocity is hardly reduced. However, when the falling water reaches the water surface its vertical velocity is suddenly reduced and it exerts a large downward force that displaces the water underneath the egg, and propels the egg upward.

6.6 Maximum height of a siphon

Demonstration
The maximum height of a water siphon is *usually* limited to around 10 meters, or one atmosphere.

Equipment
30 meters of vinyl tubing (inner diameter 1/4 inch) and two large pails.

Discussion
Raising a siphon to 10 meters is not a demo that can be conducted in most classrooms(!), but it might make a nice class project. To conduct the demo, you

need to have access to a stairwell of a building that allows you to raise a siphon at least 10 meters.

To prepare the siphon, place the two large pails at the bottom of the stairwell. It is most important to get all the air bubbles out of the tube for siphon action to begin. You can best accomplish this by using a hose or faucet to force water into the entire 30 meter length of tubing while it is all coiled up in one pail. After adding enough water to entirely fill that pail, transfer half the hose to the empty pail. Be sure to cover the end of the hose with your thumb when you transfer it, so as to keep air bubbles out of the hose. Siphon action should begin immediately, and should continue until the water levels in the pails are the same.

In order to raise the siphon to 10 meters (or higher), have someone lower a string which will be tied onto the middle of the siphon, which is to be hauled up. (To prevent the tubing from crimping, you probably should pass the tube over a pulley.) You should find that water remains in the top of the siphon until it approaches a height of 10 meters (assuming both bottom ends remain submerged in the two pails). As the top of the siphon approaches the ten meter height, you should observe that bubbles begin to form near the top, just as though the water were boiling. Once the bubbles merge at the top, and break the water column, siphon action will stop. The reason for the bubble formation is that water normally has enough disolved gases, so that very small bubbles are present. These small bubbles grow uncontrollably near the top of a 10 meter high siphon, because the hydrostatic pressure inside the liquid reaches zero at that height.

However, the 10 meter limit to water siphons is not an absolute one. For example, if one were to drive off all disolved gases by boiling the water in vacuum first, bubbles should not form until heights much greater than 10 meters, as long as there are no nucleation centers on the walls of the tube.[4, 3] In effect, you can think of the siphon as an Atwood's machine, for which the "water rope" passing over the top plays the role of the two hanging masses. The water rope idea is perfectly justified because of the extremely strong cohesive forces between water molecules, which gives a water column an extremely high tensile strength.

Effectively, at the top of a very tall siphon, the water is in the metastable condition known as "negative pressure," as it is being pulled apart by the water trying to descend in the two siphon columns. A rupture of a water column under negative pressure will eventually occur when the negative pressure is sufficient to cause cavitation—the sudden growth of microscopic bubbles. Under ideal circumstances, a water column could be raised to about 3.0 kilometers (equivalent to minus 290 Atmospheres) before breaking! Of course, no one has ever raised a siphon that high. Instead the experiments are done using a narrow Z-shaped tube of water rotated in a high-speed centrifuge about the center of the Z. The tube is rotated at an increasing speed until the water column filling the tube ruptures, and water squirts out the ends of the Z.[4].

The ability of cohesive forces between water molecules to pull a column of water above 10 meters also explains how trees are able to get water up to their tops, some of which are 110 meters above the ground. Contrary to popular belief, this ability is not due to capillary action—the force between the water molecules and the walls of the narrow capillaries in which the sap rises. Capillary action—while present—is far too small a force. In order for a tall tree to raise water up 110 meters it must exert a negative pressure of eleven atmospheres at the top end of the capillary. Evaporation of water from the leaves is the mechanism that provides this sizable negative pressure. Negative pressures of nearly three atmospheres have also been observed in octupus suckers on wetable surfaces.[5]

The idea that siphons cannot ever raise water higher than 10 meters may arise from the observation that water siphons will not work in the absence of atmospheric pressure, which you can demonstrate by placing a siphon under a bell jar from which the air is pumped out—see, for example, R. M. Graham's article.[6] In such a demo, siphon action does indeed stop once the pressure has been reduced below the vapor pressure of water. However, the same experiment can be done using a small mercury siphon instead of water. Although I was not successful when I tried it, if you use clean mercury and a clean tube, and if you succeed in driving out all small bubbles from the mercury before using it in the siphon, you should find that the siphon continues to work even after the air is pumped out of the bell jar.[7]

6.7 Narrowing of a descending water stream

Demonstration
A falling stream of water from a faucet narrows with distance in a predictable way, which allows you to calculate the flow velocity and flow rate.

Equipment
A water faucet.

Discussion
The cross section of a water stream from a faucet gets smaller, the further the stream descends. This narrowing is a direct consequence of the constancy of the volume flow rate R at all points in the stream. Volume flow rate, measured in units of volume per unit time, can be written as $R = \frac{\Delta V}{\Delta t} = A\frac{\Delta x}{\Delta t} = Av$. As the stream descends and its velocity v increases, its cross-sectional area A must inevitably decrease by the same factor.

Suppose we define y as the distance below the faucet where the diameter of the stream (initially d) is reduced to $d/2$. Obviously, after descending a distance y, the cross-sectional area is one-quarter its initial value, and hence based on the constancy of Av, its velocity is four times its initial value, that is, $v = 4v_0$. Since, falling water increases its vertical speed according to $v^2 = v_0^2 + 2gy$, so if we substitute $v = 4v_0$ in the preceding equation, we find that $v_0 = \sqrt{2gy/15}$. Thus, by observing the distance, y, for the stream diameter to be halved, we have a way of determining the speed at which the water leaves the faucet.

You might try to check this experimentally, but unfortunately initial flow velocity is not so easy to measure. One quantity that is easy to observe, however, is the amount of time for the water to fill a fixed volume such as a cup. This time is obviously just the volume of the cup divided by the volume flow rate, R, that is, $t = V/R = V/(v_0 A_0)$. Substituting $v_0 = \sqrt{2gy/15}$ and $A_0 = \pi d^2/4$ in the preceding equation yields the result

$$t = \frac{4V}{\pi d^2}\sqrt{\frac{15}{2gy}}. \tag{6.1}$$

See how close your measured time to fill a cup of volume V agrees with this prediction. When I tried the experiment using a kitchen faucet turned on part way, the stream diameter initially $d = 1.6$ cm narrowed to half that value after descending $y = 3$ cm. The time predicted to fill a measuring cup whose volume was $V = 480$ cm^3 (16 oz) is therefore 12 seconds, according to equation 6.1. The 25 percent discrepancy between the observed time (16 seconds), and the predicted time is probably not too surprising in view of the difficulty in measuring y very accurately.

6.8 Bobbing cylinder

Demonstration
If you submerge a floating cylindrical object, and let it go, it bobs up and down in simple harmonic motion at a predictable frequency.

Equipment
A bucket and a cylindrical floating object. The cylindrical object could be a glass or a glass jar, which you may need to weight with some nails in order to allow it to float in an upright orientation.

Discussion
If you submerge the cylinder slightly, and let it go, it will bob up and down a few times, until the oscillations quickly damp out. As we shall show, the predicted period of the oscillations is $T = 2\pi\sqrt{\frac{M}{\rho A g}}$, where M is the mass of the cylinder, ρ is the density of the liquid, and $A = \pi d^2/4$ is the cylinder's cross-sectional area. Try to verify this prediction by timing a few oscillations—but don't expect to obtain a precise measurement of the period, because the oscillations of the cylinder damp out rapidly. You could get a more accurate result using a much more massive cylindrical object whose oscillations should last longer.

The formula for the oscillation period is easy to derive. If a cylindrically shaped object is floating, and you release it after pressing it down by a distance x, the unbalanced upward buoyant force equals the weight of water occupying the volume Ax, namely, $F = -\rho A x g$, which by Newton's second law equals

$Ma = Md^2x/dt^2$. Therefore, we have

$$\frac{d^2x}{dt^2} = -\frac{\rho A g}{M}x = -\omega^2 x,$$

whose solution $x = x_0 \sin \omega t$ corresponds to simple harmonic motion with a frequency $\omega = \sqrt{\frac{\rho A g}{M}}$, and period $T = 2\pi/\omega = 2\pi\sqrt{\frac{M}{\rho A g}}$.

6.9 Propeller on a stick

Demonstration
By twirling a thin dowel connected to a propeller, you can generate enough lift for the stick to rise vertically in the air.

Equipment
A propeller on a stick toy.

Discussion
This toy, known since colonial days, can demonstrate some principles of aerodynamics. To prepare to launch the toy with a spin, hold the thin dowel vertically between the heel of your left palm and the fingertips of your right hand. Keep your thumbs out of the way of the propeller, and quickly move your right hand forward, rolling the dowel out to the edge of your left hand, thereby launching the spinning toy in the air. The toy should rise in the air—or at least hover a little while—assuming you spun it fast enough, and in the right direction!

How can we estimate how fast you need to spin the dowel for the toy to rise? A rough estimate can be found by calculating the volume of air swept out by the propeller each revolution, and assuming this air is given a downward velocity equal to the average tangential speed of the propeller. The tangential speed does vary from point to point on the propeller, according to $v = \omega r$, but we can take its average value to be $\omega L/4$, where L is the length of the propeller. The volume of air swept out by the propeller in the time of one revolution, $\Delta t = 2\pi/\omega$, is contained in a cylinder of volume $V = \pi L^2 h/4$, where h is the *projected* height of the propeller averaged along its length. Applying Newton's second law to find the lift force yields:

$$F = m\frac{\Delta v}{\Delta t} = \left(\frac{\rho\pi L^2 h}{4}\right)\frac{\omega L/4}{2\pi/\omega} = \frac{1}{32}\rho L^3 h\omega^2,$$

where $\rho = 1.3$ kg/m^3 is the density of air (assumed to be at rest initially). If the lift force is just able to overcome gravity ($F = mg$), we can use the preceding equation to solve for the minimum angular speed, giving

$$\omega = \sqrt{\frac{32mg}{\rho L^3 h}}.$$

If you wanted to try to check this result, you need to have some way of estimating the angular speed of the toy when it is just able to lift off. A crude estimate would be to estimate how much time it takes your hands to move past each other as you give the vertical dowel a spin. If your hands move a distance s in a time t, the angular speed they give the dowel of radius r is $\omega = s/(rt)$. A more accurate way of finding the angular speed would be to use a strobe to see how fast the propeller is turning when it leaves your hand.

But don't expect to get very close agreement with the result computed above, given the crudeness of the approximation used. When I made the measurement using my toy, the predicted angular frequency using the preceding equation was 256 rad/s. The observed strobe frequency to freeze the propeller was found to be about 1,600 flashes per minute—corresponding to 1,600 propeller revolutions per minute, or 164 rad/s—which puts my measured value within about 35 percent of the predicted value. (Note that it is very easy to mistakenly have your measured frequency off by a factor of two, because a propeller looks the same after making half a turn as it does when making one full turn, therefore put a mark on one side of the blade.)

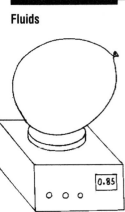

6.10 Weighing a balloon when filled and empty

Demonstration
You can find the air pressure in a balloon from its weight when it is filled and empty.

Equipment
A large (40 cm diameter) balloon, and a digital scale accurate to a hundreth of a gram, which can be found in most chemistry labs.

Discussion
When you weigh a filled balloon on a scale, the weight recorded is the difference between the actual weight and the buoyant force of the outside air. In effect, the scale records a mass m_{full} equal to the mass of the balloon plus the air it contains minus the mass of the room air displaced by the balloon. After you weigh the filled balloon, weigh it when empty, and compute $\Delta m = m_{full} - m_{empty}$, which represents the difference between the mass of the air inside the balloon and the mass of the air it displaces.

You can estimate the volume of the balloon by approximating it as a sphere. Wrap a string around it to find its circumference, and hence its radius, and then compute the volume of the balloon using the spherical approximation, $V = 4\pi r^3/3$. Finally, compute the density $\Delta\rho = \Delta m/V$, which represents the difference between the air density inside and outside the balloon. If $\Delta\rho$ is divided by the known density of air at atmospheric pressure, $\rho = 1.3$ kg/m^3, you can find the fractional or percentage difference between the air density inside and outside the balloon. Assuming the air is at the same temperature inside and out, we can write that $\frac{\Delta P}{P} = \frac{\Delta\rho}{\rho}$, that is, the fractional increase in pressure equals the fractional increase in density.

When I tried the experiment, I found $\Delta m = 0.36$ grams for a balloon of 13.7 cm radius, which gives $\Delta P/P = 0.025$, or an increase in pressure of 2.5 percent of an atmosphere inside the balloon. One very interesting aspect of the measurement is that when the balloon was first blown up and immediately put on the scale, the excess mass above the empty balloon was originally only about half as great (0.18 grams), and the value increased to about 0.36 grams over the course of a few minutes. Apparently, this effect is the result of the air from your breath cooling from body to room temperature, which results in slightly denser air in a slightly smaller balloon volume, and hence a slightly smaller buoyant force.

Notes

1. On the other hand, if you originally really filled the cup to the brim, it is possible that some overflow will eventually result if you leave the cup on the turned-on OHP long enough, because as the water temperature rises after the ice has melted, it will slowly expand over time. (Actually, since the cup's interior volume also expands as it is heated, whether or not overflow occurs depends on how much more the water expands than the cup.)

2. J. Walker, *The Flying Circus of Physics*, p. 81.

3. Chuck Warren of the Thatcher School gave me the idea for this demonstration. Neither he nor I have succeeded in efforts to raise a water siphon appreciably above a height of 10 meters.

4. Hayward, A.T.J., "Negative Pressure in Liquids: Can It Be Harnessed to Serve Man?" American Scientist, **59**, 434–43 (1971).

5. Andrew M. Smith, "Negative Pressure Generated by Octopus Suckers: A Study of the Tensile Strength of Water in Nature," the Journal of Experimental Biology, **157**, 251–71 (May 1991).

6. R. M. Graham, Journal of College Science Teaching, September/October 1995, 67–69.

7. I am indebted to Ronald Stoner for informing me that mercury does not boil in vacuum.

Chapter 7

Thermodynamics

7.1 When to add the cream to your coffee

Demonstration
Your coffee stays hot longer if you add the cream right away, rather than waiting until just before you drink it.

Equipment
An immersion heater, a digital thermometer with external probe, and a coffee mug or styrofoam cup.

Discussion
To understand why your coffee is hotter if you add the cream right away, rather than right before drinking it, we need to consider the factors that influence its rate of cooling. According to Newton's law of cooling, for small temperature differences between a body at temperature T and the room at temperature T_R, the body cools at a rate given by

$$\frac{dQ}{dt} = -k(T - T_R), \tag{7.1}$$

where k is a constant that includes the effects of conduction, convection, and radiation. Varying amounts of coffee (or water) in an uncovered cup have roughly the same exposed surface area, so we might expect k not to depend on how much coffee is in the cup. Now, a heat loss dQ can be expressed as $dQ = cmdT$, where c and m are the specific heat and mass of the coffee, and dT is its temperature drop in the time dt. Therefore, we can express equation 7.1 as

$$\frac{dT}{dt} = -E(T - T_R), \tag{7.2}$$

where $E = \frac{k}{cm}$. If equation 7.2 is integrated over time, we find

$$\Delta T = \Delta T_0 e^{-Et}, \tag{7.3}$$

where $\Delta T = T - T_R$, and $\Delta T_0 = T_0 - T_R$ are the temperature excesses over room temperature at times t and 0. Notice that this equation is obviously correct at $t = 0$ and $t = \infty$. Furthermore, since E is inversely proportional to the mass, m, equation 7.3 also agrees with our intuition that large masses cool more slowly than small ones. Essentially, the law says that the excess temperature above room temperature decreases by a constant factor during equal time intervals.

For simplicity, let us make the following assumptions:

• Coffee and cream have the same specific heat, so that water can be substituted for each of them in the demo.
• The masses of coffee and cream are equal.
• The initial temperatures are T_C for the coffee and T_R for the cream.
• You intend to drink the coffee t minutes after pouring it.
• You want the coffee to be as hot as possible when you drink it.

First, let us find the temperature in the case where you add the cream right away. If you combine equal amounts of coffee and cream, the initial temperature of the mixture will be $T_0 = \frac{1}{2}(T_R + T_C)$, midway between the individual temperatures—which is above room temperature by $\Delta T_0 = \frac{1}{2}(T_C - T_R)$. After waiting t minutes, according to Newton's law of cooling, the excess above room temperature should be reduced to

$$\Delta T_t = \frac{1}{2}(T_C - T_R)e^{-Et}. \tag{7.4}$$

We have calculated the temperature excess above room temperature when you add the cream right away (equation 7.4). Now, suppose that the cream is added just before drinking the coffee at the end of t minutes. The mass that appears in the exponent is only half what we had before, so that the new exponential factor is $E' = \frac{k}{cm/2} = 2E$. Given an initial coffee temperature of T_C, or an excess above room temperature $\Delta T_0 = T_C - T_R$, we again can find the excess above room temperature after t minutes using Newton's law of cooling, giving

$$\Delta T = \Delta T_0 e^{-E't} = (T_C - T_R)e^{-2Et}. \tag{7.5}$$

Finally, if we mix in an equal quantity of cream at room temperature, the excess temperature above room temperature is reduced in half to

$$\Delta T = \frac{1}{2}(T_C - T_R)e^{-2Et}. \tag{7.6}$$

As you can see by comparing equations 7.4 and 7.6, the temperature excess over room temperature differs by the factor e^{-Et} (a number less than one), depending on whether you add the cream right away or add it after t minutes. The basic reason that adding the cream right away keeps the coffee hot longer is that you are immediately lowering the temperature of the mixture and drastically slowing the rate of cooling as a result. By doubling the mass, you also cut down on the rate of cooling.

An experimental test

Here are some data from an experiment using 8 oz of water initially divided evenly between two uncovered styrofoam cups. First, half the water was heated to $T_C = 198°$ F, then allowed to cool for 10 minutes, and then mixed with the other half of the water which was initially at room temperature, $T_R = 62°$ F. The final temperature of the mixture was recorded as 103.3° F, after the thermometer settled down—which took about a minute. This case may be contrasted with that where the mixing occurs immediately after half the water is heated to 198° F. In this case, the temperature of the mixture after 10 minutes was 114.3° F, confirming the prediction that the mixture stays hot longer if you add the "cream" to the "coffee" right away.

How about using the results of the experiment to test the prediction from Newton's law of cooling? The table shows the temperature of half a cup of water initially heated to 198° F at successive one-minute intervals.

Time (min)	1	2	3	4	5	6	7	8	9
Temp (°F)	188.5	181.0	174.0	168.2	163.0	158.5	154.1	150.2	146.7
t-drop		9.5	7.5	7.0	5.8	5.2	4.5	4.4	3.9	3.5
E'-value		.0699	.0593	.0588	.0518	.0490	.0446	.0456	.0442	.0413

Using equations 7.2 or 7.5, we can find E given dT/dt and $T - T_R$. For example, in the first minute the hot water initially at 298° F, was found to drop by

9.5° F, so $dT/dt = -9.5°$ F per minute initially. Thus, substituting $\Delta T_0 = T_0 - T_R = 198 - 62 = 136°$ F, in equation 7.5, we find $E' = 2E = 9.5/136 = .0699$, or $E = 0.0349$. Likewise, the other entries for E' in the bottom row of the table were found in a similar manner.

An exponential variation of temperature would predict a constant value of E over time. It would, therefore, appear from the preceding data that Newton's law of cooling is true at later times, but not at early times when the water is much hotter than its surroundings—exactly when the approximation is supposed to break down. Lastly, we can also use the preceding data to find an average value of $E = 0.0258$. Using that average value, the ratio of the excess temperature above room temperature for the two mixing methods (adding the cream immediately or waiting), would therefore be predicted to be $e^{-Et} = e^{-0.258} = 0.773$. Given the measurements stated previously, the experimental result for the ratio of the excess temperatures for the two mixing methods is $(103.3 - 62) \div (114.3 - 62) = 0.800$, which is in reasonably good agreement with the predicted ratio.

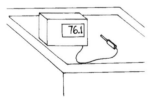

7.2 Heating black and silvered bodies

Demonstration
The rates of cooling and heating for blackened and silvered bodies can be studied using a thermometer placed on the overhead projector.

Equipment
An immersion heater, two plastic cups, and a digital thermometer with an external probe, which has been wrapped in a small piece of aluminum foil.

Discussion
In this demo we shall compare how the rates of cooling and heating for an object (the thermometer probe), depends on whether it has a blackened or silvered surface. To make the comparison, you need to in one case wrap the probe in a small piece of aluminum foil, and in the second case wrap it with a small piece of foil that has been blackened—either using a matte black spray paint or a candle flame. To

heat the probe, place it at the center of the overhead projector, and record the temperature as a function of time. You should find a dramatic difference both in the rates of heating, and also the final equilibrium temperatures for the blackened and silvered cases.

For example, after 10 minutes of heating, I found that in the unblackened case, the thermometer registered 83.0° F—only 12.4° above room temperature. But, for the blackened case, the temperature was recorded as 100.0° F after only two minutes of being placed on the OHP. The temperatures for successive two-minute intervals for the blackened case were 100.0, 114.0, 118.5, 120.7, and 121.8° F, which appears to show an asymptotic rise to some equilibrium value at which the object radiates heat at the same rate that it absorbs it.

The heating of the probe is primarily through its absorption of infrared radiation, rather than visible light. One way to show this is to place a layer of water between the surface of the OHP and the probe, since the water will largely absorb the infrared radiation, even though it is transparent to light. Fill a transparent plastic cup to a 2-centimeter depth of water, and place a second cup floating inside the first. You can then place the blackened probe at the bottom of the second cup, and observe its rise in temperature from the time it is placed there. I found a temperature rise to 107.6° F—far less than the blackened probe's rise without the water, which apparently does filter out an appreciable fraction of the infrared. The absorption of infrared radiation by water helps explain why water vapor is the planet's most important greenhouse gas.[1] (Another way to demonstrate that water absorbs the infrared radiation that causes most of the heating is mentioned in note 8 of chapter 1.)

Different rates of cooling
We can also use the blackened and silvered probe to investigate rates of cooling as well as rates of heating. Blackened objects would be expected to be much better radiators of heat, as well as much better absorbers than silvered ones. To make the comparison, you need to place the blackened and silvered probes in boiling water at the same initial temperature, and observe their respective rates of cooling once they are removed. For example, I found that after being removed from boiling water for one minute, the sil-

vered probe had cooled to 99.0° F, but that after one minute the blackened probe had cooled to 102.3° F, implying that, to my great surprise, the blackened probe cooled more slowly. Apparently, in this case the mechanism of heat loss by conduction is far more important than that of radiation, and the extra few layers of paint must reduce the heat conduction in the case of the blackened probe.

Why do we find such a difference between the heating and cooling cases, where it is the same object involved in both—the thermometer probe? For the heating case, the source of heat—the OHP—the main mechanism of heating of the probe, is through radiation, because even though the probe touches the surface of the OHP, it is not a good thermal contact, and little conduction takes place either there or through the air. In contrast, for the cooling case, it is the probe that is the source of heat, which is in good thermal contact with the foil.

There are numerous practical applications of the fact that shiny surfaces like aluminum foil make poor radiators of heat—and hence conserve the internal heat—ranging from "space blankets" to wrapping your lunch in aluminum foil. In the latter case, of course, the purpose of the foil is to keep the cold in, or equivalently, to reduce the rate of heat flow from the environment into the sandwich.

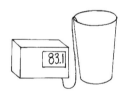

7.3 Heating by convection versus conduction

Demonstration
For fluids, such as water, convection is a much more efficient heat transfer mechanism than conduction.

Equipment
An immersion heater, a large (16 oz) styrofoam cup, a graduated cylinder, and a digital thermometer with an external probe.

Discussion
If a large cup of water is heated using an immersion heater placed near the bottom of the cup, a convection current will be created as the heated water rises, and the colder water sinks. However, if the immersion heater is placed just below the water surface, the

heated (less dense) water will not sink, and no convection current will be created. Thus, heating the top layers of water results primarily in heating via conduction, while heating the lower layers results in heating via convection. If the thermometer probe is placed near the bottom of the cup away from the heater, you will find radically different variations in temperature with time in the two cases.

For example, after 5 minutes of heating, I found that when the immersion heater was near the water surface, the temperature near the bottom had risen a mere $0.7°$ F. In contrast, when the heater was on the bottom on the opposite side from the probe, the temperature was found to rise $52.7°$ in 5 minutes, and even more at the top of the water. Given this $70:1$ ratio in the two rises in temperature over a five-minute period, what might you expect for the difference in boiling times for the two heating methods?

Surprisingly, the time for the water in the cup to be brought to a vigorous boil is less than 50 percent different for the two cases: 13 versus 18 minutes. How can this relatively small difference be explained? When the heater is just below the water surface, the temperature at the bottom initially rises very slowly, as a very hot layer of water heated by conduction gradually descends. In other words, at any given time, the variation of temperature with depth, or "temperature profile," has a very sharp gradient at a certain depth—a depth which slowly descends as more and more water is heated. As a result, the temperature of the water at the bottom suddenly rises during the last few minutes of heating—more than half the temperature rise occurred during the last 10 percent of the time.

Even though heating by conduction is much less efficient than heating by convection, the two methods would, in fact, give identical times to boiling were it not for heat loss to the environment—a consequence of the conservation of energy law. Thus, while the total amount of heat added to the water is the same for both heating methods, the temperature profile is drastically different in the two cases, and it serves as a better measure of the relative inefficiency of heating by conduction.

Here is a simple way to demonstrate that hot water stays on top without mixing with the colder water below. Fill a transparent graduated cylinder with cold water. Pour half the water into a cup, add some food coloring to the water, and then heat the colored water in the cup using an immersion heater. Gently pour the hot colored water back into the cylinder containing cold water, by tilting the cylinder so that you pour the water along the side. You should find that after the transfer, the lower portion of the water remains clear, showing that the hot dyed water stays in the upper portion of the cylinder. However, don't expect to find a sharp boundary between the dyed and clear water, because some mixing is inevitable during the water transfer.

Notes

1. Water vapor, the dominant greenhouse gas, has allowed life to flourish on the planet by warming its surface above subfreezing temperatures that would prevail in the absence of a natural greenhouse effect. The main concern about global warming is with additions to the greenhouse gases through human activities, specifically by-products like carbon dioxide and methane.

Chapter 8

Mechanical Oscillations and Waves

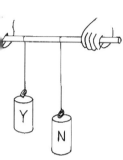

8.1 "Yes/no" pendulums

Demonstration
When you subtly move a bar from which a pair of different length pendulums hang, it is possible to excite only one of the "yes/no" pendulums into large amplitude oscillations that seem to the audience to arise from no apparent physical cause.

Equipment
A pair of pendulums consisting of aluminum soda cans tied to strings hanging from a bar. Make the pendulum lengths differ by around 20 percent, and add enough water to each can to fill it about a quarter full. For effect, you may wish to label the cans "Yes" and "No."

Discussion
This humorous demo—a spoof of "psychic powers"—makes a serious point about resonance. When one of the cans is driven resonantly by shaking the bar very slightly (without the audience being able to perceive the motion), the hanging cans seem to answer yes/no questions asked by you or the audience. (You might want to include a third "maybe" can, between the other two.)

With a little practice, you should find it quite easy to excite only one can into large oscillations by moving the supporting bar so imperceptibly that the can seems to start oscillating completely spontaneously. For best control of the motion of the bar, you may wish to sit down, and hold the bar in two hands, while resting your elbows on your knees. The point of adding some water to the cans is to give you some kinesthetic feedback on how to move the bar to achieve resonance. When you get proficient at exciting only one can by almost imperceptible movement of the bar, you almost may convince yourself that one of the cans is responding to your telekinetic powers!

In addition to making an important point about resonance (namely, that large oscillations can result from a very small amplitude of the source if the frequency and phase are right), the demo also may say something about unexplained phenomena that seem to arise with no (apparent) physical cause. Whether or not you believe that ESP or other related topics have any basis in fact, it is worth pointing out to the audience that very small external influences (perhaps even unconscious ones), can, through resonance, have a large effect on a system.

8.2 Swinging your arms while walking

Demonstration
The "natural" period with which you swing your arms while walking is that of a physical pendulum of the same length.

Equipment
None.

Discussion
When you swing your arms naturally while walking, the period of the swings is around a second. You can show an audience how strange it would look if you were to swing your arms much faster or slower than the natural frequency. To check how close the natural frequency is to what would be expected for a physical pendulum, have someone time the period for some number of full (back and forth) swings as you walk, while swinging your arms naturally. Compare the measured period with that predicted for a physical pendulum having the length of your arm: $T = 1.64\sqrt{\ell}$, where ℓ is the length of your arm in meters. This formula, derived in the next section, unrealistically assumes that your arm has a uniform cross section, and pivots about a point at its top end—so don't expect very close agreement.

Using a similar sort of argument, you can figure out the natural human walking speed, by treating your swinging leg as a physical pendulum. For example, assuming your leg has a length of one meter, it will therefore make a complete swing in 1.64 seconds, and a half-swing in $0.82 \approx 1$ seconds. Assuming that your foot moves through an arc of perhaps 60 degrees in that time, it travels a horizontal distance of one meter, yielding a typical walking speed of approximately 1 m/s.[1]

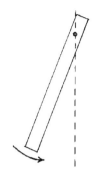

8.3 Period of a physical pendulum

Demonstration
Make a physical pendulum using a meter stick, and observe how its period depends on the distance from the end of the ruler to the pivot point.

Equipment
A meter stick.

Discussion
Pendulums come in many variations from the simple pendulum consisting of a mass m on a string of length ℓ to the physical pendulum you might find as the swinging bar in a grandfather clock. For simplicity, let us suppose the physical pendulum has a uniform cross section, and that it swings freely about a point close to its top end. We might make such a pendulum by drilling a small hole in a meter stick a few millimeters from its end, and suspending it from a straightened paper clip. As we will later show, the period of such a pendulum can be expressed as $T = 2\pi\sqrt{2\ell/3g}$, which in the last demo we wrote as $T = 1.64\sqrt{\ell}$, using $g = 9.8\,\text{m/s}^2$. Notice that the period of a swinging meter stick ($\ell = 1\,\text{m}$) is predicted to be 1.64 seconds, which you could check by timing some number of full swings.

But here is a more interesting test. Make a simple pendulum (mass on a string), of length $\ell = 2/3$ meter, whose period is therefore expected to be $T = 2\pi\sqrt{\ell/g} = 2\pi\sqrt{2/3g}$. Based on the preceding equation for the period of a physical pendulum, this simple pendulum of length 2/3 meter should have exactly the same period as a physical pendulum of length one meter. All you need to do to verify this prediction is to allow the mass on the string of length 2/3 meter swing side by side with the meter stick on the paper clip, and see whether they remain swinging together.

Another interesting variation of this demo is to allow the meter stick to swing about some point other than one next to one end. How do you suppose the period might change if the pivot point were made nearer the center of the stick? The general equation for the period of a physical pendulum with an arbitrary pivot point can be expressed as $T = 2\pi\sqrt{I/mgd}$, where I is the moment of inertia, m is the mass, and d is the distance of the pivot to the center of gravity. Notice that this equation predicts an infinite period when $d = 0$ (pivot at the center of the stick), which is reasonable, because in that case the stick would not swing at all.

If the period is infinite for a pivot point right at the center of the stick, you might think that the period would gradually lengthen as the pivot moved from one end toward the center, but surprisingly that is not the case. In fact, the preceding equation can be used to show that as the pivot approaches the center from one end, the period actually decreases for a while before eventually increasing to infinity. As a result, there is a pivot point on the stick located a sixth of a meter from the center ($d = 1/6$ meter), which has exactly the same period as a point right next to the end. You could easily verify this prediction by using two meter sticks side by side, and seeing if they swing together when one pivots about a point at one end, while the other pivots about the point 1/6 meter from the center.

It is not difficult to prove the last result. We need a formula for the moment of inertia of the meter stick, which can be found using the parallel axis theorem: $I = I_{CM} + md^2$. If the axis is 1/6 meter from the center, we use $d_1 = \ell/6$ and $I_{CM} = m\ell^2/12$, to find that $I_1 = m\ell^2/9$, and hence, we obtain $T_1 = 2\pi\sqrt{I_1/mgd_1} = 2\pi\sqrt{2\ell/3g}$. But, if the axis is right at one end, we use $d_2 = \ell/2$ and $I_{CM} = m\ell^2/12$, to find that $I_2 = m\ell^2/3$, and hence, we obtain $T_2 = 2\pi\sqrt{I_2/mgd_2} = 2\pi\sqrt{2\ell/3g}$—exactly the same result. Therefore, the periods for pivot points at one end and 1/6 meter from the center must be identical.

8.4 Oscillations of a ruler on a cylinder

Demonstration
A ruler balanced horizontally with its midpoint resting on a cylinder will oscillate with a specific frequency.

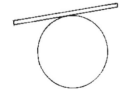

Equipment
A soda can and a 12-inch (30 cm) ruler.

Discussion
Balance a ruler horizontally atop a cylinder, such as a soda can. When the ruler is given a push away from its equilibrium horizontal orientation, it undergoes simple harmonic oscillations. It can be shown that, for small oscillations, the period can be expressed as

$$T = 2\pi \sqrt{\frac{\ell^2}{12g(r - d)}}, \tag{8.1}$$

where ℓ is the length of the ruler, r is the radius of the cylinder, and d is the very small distance the ruler's center of mass lies above the contact point of the ruler with the cylinder. (For a ruler of rectangular cross section, d would equal half the thickness of the ruler.) You can test the validity of this equation by observing the oscillation periods using cylinders having various radii.

If you try the experiment using cylinders having radii that are not much larger than d, such as stick pens or pencils having a round cross section, you will probably find that achieving balance is difficult, and that only very small oscillations can be observed. For cylinders having radii less than d, the equilibrium is unstable, and no oscillations can be observed. Notice that the equation 8.1 gives an imaginary result for $d > r$, because in this case the torque due to the weight of the ruler drives the ruler away from equilibrium rather than toward it.

To derive equation 8.1, we need an expression for the restoring torque around the contact point due to the force of gravity (acting at the ruler's center). This torque has a moment arm given by $r_\perp = (r - d)\sin\theta$, so that by Newton's second law $\tau = -mgr_\perp = Id^2\theta/dt^2$. For small oscillations the contact point is never very far from the center of the ruler, therefore we may approximate the moment of inertia about the contact point by its center of mass value, $I = m\ell^2/12$. After we substitute this result in Newton's second law, and use the small angle approximation $\sin\theta \approx \theta$, we have $d^2\theta/dt^2 = -\omega^2\theta$, where $\omega = \sqrt{12g(r - d)/\ell^2}$. If $r > d$, the solution of this differential equation corre-

sponds to oscillations with the period given by equation 8.1.

8.5 The partial ring pendulum

Demonstration
Pendulums made from partial rings having a common radius oscillate with identical periods, independent of the fraction of the ring.

Equipment
Several plastic rings of the type that used to be inserted into magnetic tapes, which can be obtained from most computer centers. You could also use embroidery hoops.

Discussion
Cut several arcs from the rings corresponding to 1/4, 1/2, and 3/4 of a ring. Together with one uncut complete ring, this gives you a set of four partial rings—if we lump a complete ring into the "partial" category. We wish to observe oscillations of each partial ring about an axis perpendicular to the plane of the ring. To suspend each ring from an axis, make a small hole in the center of each arc, and support each one by a straightened paper clip through the hole. Obviously, the hole needs to be slightly larger than the paper clip, so the ring can swing with little friction. You may need to resonantly drive the swings a little to keep the oscillations going.

You also may wish to put two partial rings on the same paper clip, so that you can observe if they oscillate side by side with a common period. The period is predicted to be $T = 2\pi\sqrt{2R/g}$, where R is the radius of the ring. (Notice that there is predicted to be no dependence of period on the arc length of the partial ring.) Not surprisingly, the period for a partial ring of infinite radius (a straight stick free to pivot about an axis through its center) is predicted to be infinite, because in such a case the center of gravity is located right at the axis.

Although the independence of period on amount of arc is theoretically true, as shown by David Wagner, Thomas Walkiewicz, and David Giltinan,[2] very small differences in period can occur, depending for example, on having a radial offset of the support hole.

Nevertheless, you are unlikely to notice any difference in the periods of the various partial and complete rings, to within a few percent accuracy, with the possible exception of the 1/4 ring.

Is there some obvious reason why the period of a partial ring is independent of the arc? In general, the period of a physical pendulum can be expressed as $T = 2\pi\sqrt{I/mgd}$, where I is the moment of inertia, mg is the weight, and d is the distance the center of mass lies below the axis. If we apply this formula to a partial ring, it turns out that both I and d have the same dependence on the amount of arc, so that the period itself does not depend on the arc.

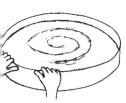

8.6 Rotating waves

Demonstration
Rotating waves can easily be generated in a circular water tank on an OHP.

Equipment
A transparent flexible circular water container, such as a 12-inch (30 cm)-diameter dish of the type placed underneath a large potted plant, or a large flexible plastic cake cover.

Discussion
Usually we think of waves as being either of the standing or traveling type, but rotating waves represent a third major category. Briefly, a rotating wave travels in circles "chasing its own tail" in a swirling motion. More details on rotating waves can be found in an article by Peter Ceperley, on which this demo is based.[3] It is quite easy to generate various rotating wave modes using a circular tank filled to a depth of about 5 cm. To excite the wave, grasp the sides of the flexible container with both hands located about 10 cm apart, and rhythmically push the tank sides.[4]

You will need to adjust both the phase and frequency of your two hand pushes to excite a particular mode. In general, you want the phase difference between the pushes to correspond to the time a rotating wave would travel along the arc between your hands. The period of your pushes needs to equal the time the rotating wave completes one revolution in order to achieve resonance. Rotating waves are important in various physical situations where waves have angular momentum, such as the quantum mechanical wave function of an atom.[5] A classical application of rotating waves would be the rotating "spiral density wave" that apparently explains the stability of spiral arms in our galaxy.

8.7 Resonant vibrations of a ruler

Demonstration
A plastic ruler held against a board and plucked creates the greatest amount of resonant or "sympathetic" vibrations at the other end when held at its midpoint.

Equipment
A plastic ruler and a board.

Discussion
The natural frequency of oscillations of a ruler clamped at one end depends on the length of overhang, with higher frequencies corresponding to shorter lengths. If the ruler is clamped in the middle (equal length overhangs), both ends have the same natural frequency, and if you pluck one end, the other end vibrates "sympathetically." To illustrate this idea, place a board having a thickness of about 2 cm perpendicular to a firm surface such as a table. Balance a horizontal ruler with its midpoint resting on the edge of the board. Press your thumb down very firmly on the center of the ruler, and pluck one end. Notice that the other (free) end vibrates with large amplitude. These resonant oscillations occur because the vibrations of the plucked end cause small vibrations in the supporting board, which resonantly drive the free end. Now repeat the demonstration, but with the ruler held down firmly at a point other than its midpoint.

The amplitude of oscillations at the free end will be smaller than before, because the frequency of the "driving" oscillations of the point of support is no longer matched to the natural frequency of the free end. Obviously, the closer the point of support is to the midpoint of the ruler, the smaller the mismatch in frequencies, and the greater the amplitude in oscillations of the free end. A simple modification of the demonstration allows you to perform it on an overhead projector. Press the midpoint of the ruler against the vertical column that supports the projector lens, and the ruler vibrations will occur in a horizontal plane, making them readily visible on the screen.

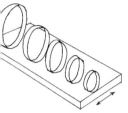

8.8 Resonant rings

Demonstration

When you shake a base containing five circular paper rings having different diameters, the frequency of shaking determines which ring vibrates with the greatest amplitude.

Equipment

A piece of stiff cardboard and some construction paper. If you want to do the demonstration on the OHP, use a piece of lucite or blank overhead transparency sheet instead of the cardboard to serve as the base on which to tape the paper rings made from the construction paper.

Discussion

Make the rings by cutting five strips of construction paper of width several centimeters, and lengths $\ell_1 = 15$, $\ell_2 = 20$, $\ell_3 = 25$, $\ell_4 = 30$, and $\ell_5 = 35$ cm. Tape the two ends of each strip to make a circular ring, and then tape each ring to the cardboard or transparent base, so that the rings stand vertically next to one another, in order from smallest to largest.

If you shake the base from side to side (in the plane of the rings), the frequency of your shaking will probably cause one ring to vibrate more than the others, depending on the frequency of your shaking. A maximum amplitude (resonance) occurs when your frequency of shaking matches the natural frequency of (free) oscillations of a particular ring. Since the natural frequency decreases the larger the ring, you should find that if you slowly increase the frequency of your shaking, each ring in sequence starting with the largest one will go through a resonance. You may find that making a chosen ring resonate takes a little hand-eye coordination, and that a small amplitude of shaking (at the right frequency) is more effective in creating a particular resonance.

You might want to use a metronome to help you keep the proper beat when exciting a particular ring to resonate. More important, the metronome will also allow you to measure each ring's resonant frequency. Although I know of no simple formula here to predict the resonant frequencies, we might expect that they should scale according to some power law, with the frequencies being proportional to some negative power of the radii of the rings, or of the lengths, that is, $f = C\ell^{-N}$. We can eliminate the unknown constant C by expressing the kth frequency as a ratio of the first one, that is, $f_k/f_1 = (\ell_1/\ell_k)^N$. Finally, we can solve the preceding equation for the unknown exponent N:

$$N = \frac{\ln(f_k/f_1)}{\ln(\ell_1/\ell_k)}.$$

We could apply the preceding equation to each of the measured frequencies f_k, for $k = 2, 3, 4, 5$. The hypothesis of a power law would be validated if a consistent exponent N is found for each k. (Note that, depending on the thickness of your construction paper, you may need to use a different set of five paper lengths if the highest frequency is above that of your metronome.) This demonstration is a modified version of the "Resonant Rings" experiment in *The Exploratorium Science Snackbook*.[6]

8.9 Ball in a parabolic potential well

Demonstration
A ball oscillating in a parabolic potential well oscillates with a specific period.

Equipment
Two grooved rulers, a one-inch (2.54 cm)-diameter stainless steel ball, and some sponges.

Discussion
To make a potential well having an (approximately) parabolic shape, put one ruler on top of a second, and tape them together at the middle. (Both should

have their grooved sides facing up.) Pry the ends of the rulers apart, and wedge a 2-cm-thick piece of sponge or foam rubber in between the rulers at each end, causing each ruler to flex into a concave-outward shape. Remove any tape that obstructs the groove on the top ruler.

Place the two rulers on a horizontal surface, such as an overhead projector. When a smooth metal ball is rolled in the groove of the top ruler, it will roll back and forth between turning points equidistant from the center, with the oscillations persisting for a considerable time. Count a large number of oscillations, and see how well the period matches the small amplitude prediction

$$T = 2\pi\ell\sqrt{\frac{7}{10gd}}, \tag{8.2}$$

where ℓ is the length of the ruler, g is the acceleration of gravity, and d is the separation of the ends of the ruler caused by the insertion of the sponge. In addition to observing free oscillations, you can also demonstrate driven oscillations and resonance by very gently moving the rulers back and forth, at the proper frequency and phase.

The remainder of this section will consist of a derivation of equation 8.2. We start with the formula for the period of a simple pendulum: $T = 2\pi\sqrt{\frac{\ell}{g}}$. A frictionless mass sliding in a well of radius of curvature $r = \ell$ has exactly the same dynamics as a pendulum at the end of a string of length ℓ. However, the use of a rolling solid ball instead of a sliding mass effectively changes the value of g to $5g/7$, because the center of mass of the ball does not accelerate as fast as a frictionless sliding mass in view of its moment of inertia. Thus, we may write for the period of a ball rolling in a parabolic well: $T = 2\pi\sqrt{\frac{7r}{5g}}$. For a circular arc, the radius of curvature, r, can be found from the approximate geometrical relation $r = \ell^2/2d$. (Even though the exact shape of a bent ruler is closer to parabolic than circular, the approximation is not too bad in the present case, where the bending is small.) Substituting $r = \ell^2/2d$ into the previous equation for the period yields equation 8.2.

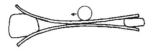

8.10 Ball in an asymmetric potential well

Demonstration
A ball oscillating in an asymmetric potential well can be used to model thermal expansion (of most materials).

Equipment
Two grooved transparent plastic rulers, a one-inch (2.54 cm)-diameter stainless steel ball, and some sponges.

Discussion
The potential energy between molecules in a solid can be expressed as a function of the intermolecular separation, r. The shape of such potential wells is usually asymmetric about the bottom of the well, with the potential energy rising more steeply for small r than large r. To make an asymmetric potential well, glue or tape two transparent rulers at their center (being careful not to obstruct the ruler groove), and insert different thicknesses of sponge between the rulers at the two ends—using perhaps twice as much sponge at the left end as the right (see figure). Place the ball in the top ruler groove, and give it a push, causing the ball to oscillate.

Observe the end points of each oscillation. Due to the asymmetry of the well, the right turning point (where the well rises less steeply), should be twice as far from the ruler center as the left one. Thus, if you observe the ball's oscillations over time as they gradually damp out, the midpoint of the oscillations should gradually move toward the bottom of the well (the center of the ruler). The idea is quite analogous to what happens as a material cools, and atomic oscillations become less vigorous. Such cooling also causes the midpoint of the oscillations to shift toward smaller values, resulting in contraction of the material.

A small fraction of materials (including water) contract when heated over some temperature range.[7] In this case, the asymmetry in the shape of the potential well would be opposite of what we have considered here—a more gradual rise on the small r side of the well than the large r side.

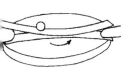

8.11 Ball in a rotating single well potential

Demonstration
A ball can oscillate in a rotating parabolic well only up to a certain critical angular velocity.

Equipment
A one-inch (2.54 cm)-diameter stainless steel ball, a turntable, such as a "lazy Susan," two grooved transparent plastic rulers, and some sponges. (If you want to show this demonstration on an overhead projector, you can make a transparent low-friction turntable using a ball bearing sandwiched between two lucite disks.)

Discussion
Make a parabolic well potential using two rulers, as described in demonstration 8.9. Place the two rulers on the rotatable turntable, with their centers coincident with the center of the turntable. Place the ball on the groove of the ruler, a distance x from the axis, and rotate the turntable. If the turntable is rotated at a certain critical angular velocity, ω, the centripetal force $m\omega^2 x$ just matches the component of gravity along the incline, $mg\tan\theta$, where θ is the angle of the ruler with the horizontal at a distance x from the center. If the curved ruler shape is parabolic (so that $y \propto x^2$, and hence $dy/dx = \tan\theta$ is proportional to x), the match between centripetal force and gravity along the incline will hold at all x.

Similarly, if the centripetal force is greater than or less than the component of gravity along the incline, that imbalance will have the same sign at all points. As a consequence, a stationary ball at an arbitrary point in the groove should either move toward the center or away from it, depending on whether the ruler is rotated at less or more than a specific rotation rate. To find the critical rotation rate, we need an expression for the shape of the parabolic well. The proper quadratic form is evidently $y = 2d(x/\ell)^2$, since at the ends of the ruler ($x = \pm\ell/2$) we find that $y = d/2$, or half the separation between the two rulers. Thus, we find that $y' = \tan\theta = 4dx/\ell^2$. Finally, setting $mg\tan\theta = m\omega^2 x$ yields the critical angular velocity $\omega = \sqrt{4gd/\ell^2}$ rad/sec.

If you wish to find the critical angular speed experimentally, you need to rotate the turntable (by hand) at a slowly increasing speed, and see when a ball in the groove starts moving outward. When a ball does begin to fly outward you need to decrease the rotation speed slightly. In practice, continual speed adjustments may be necessary to keep the ball moving back and forth between the center and end of the ruler. You could then have someone measure the time for a specific number of rotations to find the angular speed, and see if it matches the predicted value. If your turntable is quite friction-free, you can observe that as you approach the critical angular velocity from below, the period of oscillations of the ball steadily increases. (Right at the critical angular velocity the ball can theoretically remain at rest at any point on the ruler.)

You might wonder whether a better way to do the demonstration would be to use a rotating turntable of known speed, such as a $33\frac{1}{3}$ rpm record turntable, and make the curvature of the ruler what it needs to be to match that particular speed. However, one problem with this alternative is that, in practice, the ruler curvature will either be slightly smaller or slightly larger than it needs to be for the particular speed. In the former case, the ball will stay at the center (or oscillate very slowly about the center), and in the latter case, it will move outward, so a controlled *variable* speed turntable is probably required.

8.12 Stability of an inverted pendulum

Demonstration
The inverted pendulum—normally unstable—can be stabilized if the axis is oscillated in an appropriate manner.

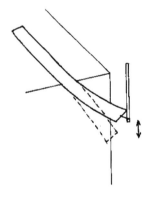

Equipment
A plastic ruler and a light piece of wood, such as a coffee stirrer or popsicle stick.

Discussion
A light physical pendulum can be made from the popsicle stick if you make a hole at one end of the stick, and place a thin nail through the hole to serve as an axle for the pendulum. The role of the plastic ruler is

to provide an oscillating point of support for the pendulum. Press one end of the ruler down at the edge of a table, so that the ruler projects forward like a diving board with the pendulum at its end. Now tape the nail to the end of the ruler, so as to allow the pendulum to rotate freely in a vertical circle. If you put the pendulum in the inverted position, and release it, it will, of course, topple over. However, if the end of the ruler is plucked, with the pendulum initially in the inverted position, it will not topple over for a few seconds, during which the ruler vibrates with sufficient amplitude. How can we explain this strange behavior?

Consider the torque on the inverted pendulum when it is a small angle θ away from its inverted orientation: $\tau = +\frac{1}{2}mgl\theta$, where the plus sign reminds us that this is not a restoring torque. In the absence of ruler vibrations, Newton's second law—$Id^2\theta/dt^2 = +\frac{1}{2}mgl\theta$—yields only divergent solutions involving exponential or hyperbolic functions of time. But, now suppose the ruler vibrates with a frequency ω_0, so that the end of the ruler has an instantaneous acceleration $a \sin \omega_0 t$. In the noninertial reference frame moving with the end of the ruler, we may take the acceleration of gravity to be $g + a \sin \omega_0 t$, so that Newton's second law becomes

$$I\frac{d^2\theta}{dt^2} = +\frac{1}{2}ml\theta(g + a \sin \omega_0 t). \tag{8.3}$$

Note that when $a > g$, the torque on the pendulum is of the restoring type (negative sign) for part of each ruler vibration. (A restoring torque occurs at any time t for which $\sin \omega_0 t < -g/a$, leading to a negative right-hand side of equation 8.3.) If the torque is of the restoring type for a sufficiently large fraction of each vibration (that is, if a is large enough), then equation 8.3 admits an oscillatory solution about the vertical, and the inverted ruler will not topple. However, when the ruler vibrations gradually damp out, the fraction of each oscillation over which the sign of the torque is negative decreases, and eventually the inverted ruler becomes unstable, and it topples.

There is still another reason that partly explains the stability of an inverted pendulum when the point of support oscillates. In practice, the end of the ruler moves in an arc of a circle, rather than straight up and down. Consequently, the pendulum rocks back and forth in a direction at right angles to the plane of its swings. This rocking motion causes the friction at the axle to increase, and helps stabilize the pendulum in the inverted position.

8.13 Finding the mass of air inside a balloon

Demonstration
The mass of air inside a large balloon can be deduced based on the frequency of oscillations if the balloon is put into resonant oscillations at the end of a rubber band.

Equipment
A large (40 cm diameter) balloon, a thin rubber band, and a weight in the range of 40 to 70 grams. The weight might be a one-inch (2.54 cm)-diameter steel ball.

Discussion
It is not so easy to find directly the mass of air that a balloon contains. For example, the most direct method of weighing the balloon while filled and empty will not work, because the filled balloon experiences an upward buoyant force that almost cancels out the weight of the air it contains. (The two forces would cancel exactly if the air inside the balloon were at atmospheric pressure instead of being slightly higher.) One way to measure the mass of any object is to put it into simple harmonic motion, and measure the oscillation period.

A mass m hung on a spring of force constant k will oscillate freely with a period $T = 2\pi\sqrt{m/k}$. Here it is simplest to use a rubber band rather than a spring. Even though rubber bands don't obey Hooke's law over a large range, the oscillations will be simple harmonic if their amplitude is kept small. Of course, if you hang a balloon at the end of a rubber band, you won't be able to cause it to go through a complete oscillation unless you hang some additional mass on the bottom of the balloon, so as to keep the rubber band under tension for the entire oscillation. One way to attach the mass would be to tape it to the bottom of the balloon, or alternatively, if you used a one-inch-diameter steel ball you could, with a bit of difficulty, force the ball through the neck of the balloon before it is blown up—though this method of adding the weight may result in some popped balloons during the experiment.

Once the weight has been added to the balloon, set it into small oscillations as it hangs from the end of the rubber band, by gently shaking the top end of the rubber band up and down at the resonant frequency. In order to observe the period, T_1, of resonant oscillations for the balloon plus the extra mass, you will need to count a large number of them during the course of a minute or two. Later, after the balloon has had its diameter measured, and has been deflated, you will also want to observe the oscillation period, T_2, when the mass is hung on the deflated balloon—again measuring the time for many oscillations, in order to get an accurate period. (Be sure that you use the same length of rubber band in both cases.)

The ratio of the two periods is predicted to be

$$T_1/T_2 = \frac{2\pi\sqrt{(m+M)/k}}{2\pi\sqrt{M/k}} = \sqrt{m/M + 1}$$

where M is the mass of the deflated balloon plus any added mass, and m is the mass of air inside the inflated balloon. Solving for m we find

$$m = M[(T_1/T_2)^2 - 1].$$

Thus, based on the measured periods with the balloon inflated and deflated, T_1 and T_2, we can compute the mass of the air inside the balloon using the preceding equation. You could then find the density of air by dividing the mass by the volume of the balloon, which you could approximate as a sphere. When I did the experiment using a one-inch-diameter steel ball as the added mass, I found that the balloon plus mass oscillated 72 ± 1 times in a minute when deflated, and 60 ± 1 times in a minute when inflated. These measurements give periods $T_1 = 1.00 \pm 0.016$ sec and $T_2 = 0.833 \pm 0.011$ sec. Using the preceding equation with $M = 0.067\,\text{kg}$, for the steel ball plus balloon, we find that the mass of air inside the balloon is $m = 0.029 \pm 0.006$ kg.

Since the balloon's diameter was measured to be 0.32 meters, this mass of air yields a value for the density of air inside the balloon of 1.8 ± 0.4 kg/m^3, where the only source of measurement uncertainty used has been that due to the measured periods. Surprisingly, the uncertainty in the periods produces more uncertainty in the final result than any uncertainty associated with the volume of the irregularly shaped balloon, even though the periods are individually measured far more accurately than the volume. A more precise final result would have been obtained had a larger balloon been used, and had the added mass been smaller in relation to the mass of the air, resulting in two periods T_1 and T_2 that were more unequal.

Could the small departure of the measured density of air from the standard value (1.3 kg/m^3), arise because the air inside the balloon is more dense than the outside air, since its pressure is higher? As shown in demo 6.10, the increased density and pressure is only about 2.5 percent, and it is too small to explain the difference, which is probably due either to random measurement error or our neglect of air viscosity.

8.14 Speed of sound

Demonstration
You can measure the speed of sound by rhythmically tapping at a frequency such that the echo is *not* heard from a distant wall.

Equipment
A metronome, a hammer, and a metal pipe.

Discussion
The first of two methods for measuring the speed of sound discussed below is based on echoes. Find an outdoor location where you can receive echoes bounced off a building at a distance d around 80 meters, with few other sources of echoes around. Test the audibility of the echoes by tapping on the metal pipe with the hammer. Now, start the metronome, and rhythmically tap the pipe to coincide with the metronome beats. In general, you should hear the

echoes coming between the taps themselves. However, you should also be able to find some lowest metronome setting (and corresponding tapping frequency f), where the echoes are not heard, because each echo arrives at the same time as the next tap.

In this case, the round-trip travel time of the sound is $t = 1/f = 2d/v$, where v is the speed of sound, which can therefore be found in terms of measured quantities from $v = 2df$. (Of course, you do need to be sure that the echoes are arriving back at your ear in the elapsed time between consecutive taps, so you should be sure that the observed frequency is the lowest one for which no echo is heard. Given that the speed of sound is around 330 m/s, the metronome frequency setting at a distance $d = 80$ meters from the wall should be around 2.1 beats per second or 124 beats per minute.)

The preceding method for finding the speed of sound is a variation of that described by George Biehl[8]. Another version of the method dispenses with the need for echoes, but does require that you have a friend located a measured distance (about 100 meters) away. While you tap the pipe with the hammer at a steady rhythm matching the metronome frequency f, your friend should change his distance d from you until he observes that you strike the pipe at the exact moment he hears the sound. He is, of course, hearing the sound from the previous tap, so the speed of sound can be found from $v = d/t = df$.

8.15 Doppler effect and shock waves

Demonstration
You can observe the nonconcentric expanding circular wave fronts associated with the Doppler effect or shock waves if you dunk a stick with many pins in a tray of water.

Equipment
A water tray at least 40 cm long, and a 30-cm-long stick with square cross section (about 1 cm by 1 cm), in which you have inserted 6 to 10 round-headed pins equally spaced by about 2 cm all along the stick starting from one end. The pins should all be inserted to the same depth.

Mechanical Oscillations and Waves

Discussion

Add water to the tray to a depth of several centimeters, and place it on the OHP. If you hold the stick horizontally just above the water surface, and dunk it so the pins all enter the water the same moment, you will produce a number of expanding circular wave fronts with centers 2 cm apart. Now, suppose that you hold the stick at a small angle with the horizontal, and move it vertically into the water, so the pin heads enter the water sequentially. In this case, you will mimic the situation of a moving source emitting pulses.

The effective source velocity v_s relative to the wave velocity c depends on the angle the stick makes with the horizontal as it enters the water, and also the speed that it descends vertically. The effective source velocity is given by $v_s = v \cot \theta$, where v is the actual velocity that you move the stick downward, and θ is its angle with the horizontal. Notice that based on the preceding equation, high source velocities correspond to smaller angles with the horizontal, or more rapid descents of the stick. (The case of a horizontal stick obviously corresponds to an infinite source velocity, since all the pins enter the water simultaneously.) You should find that if you keep the speed that the stick enters the water roughly constant from trial to trial, very small angles with the horizontal correspond to shock wave patterns ($v_s > c$), while larger angles correspond to Doppler patterns ($v_s < c$.) The difference in the two types of patterns is that in the Doppler case, the larger wavefronts completely enclose the smaller ones. For example, the figure shows a shock wave not a Doppler pattern.

8.16 Beats using a tuning fork

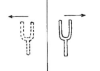

Demonstration

If you hold a vibrating tuning fork while walking away from a sound reflecting wall, listeners will hear beats. For best results, use a tuning fork mounted on a wooden box, which acts as a resonant cavity.

Equipment

A tuning fork.

Discussion

The frequency of sound from a moving source will be Doppler shifted according to a listener. For low velocities of the source v compared to the speed of sound c, the fractional increase in frequency of the source perceived by the listener is $\frac{\Delta f}{f} = \pm v/c$, where the plus and minus signs hold when the source moves directly toward or away from the listener. Equivalently, we can write the Doppler shifted frequency as $f_\pm = f(1 \pm v/c)$. If you stand in front of a wall while holding a tuning fork, a listener in the room hears sound coming directly from the fork as well as sound reflected off the wall behind it. The reflected sound appears to come from an "image" source an equal distance behind the wall.

Therefore, if you walk away from the wall at speed v approaching the listener, the image source is moving in the opposite direction at the same speed. For low velocities, the Doppler shifted frequencies of the source and its image are therefore $f_1 = f(1 + v/c)$ and $f_2 = f(1 - v/c)$. A listener in the room will hear beats corresponding to the difference frequency $f_1 - f_2 = 2fv/c$. How many beats per second might be expected using a tuning fork of frequency, say, 440 Hz? A fast walk is roughly 0.5 percent the speed of sound ($v/c = .005$), so the predicted beat frequency would be 4.4 Hz, which should be quite easy to hear.

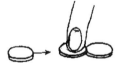

8.17 Three-cent shock waves

Demonstration

The shock wave created when one penny is flicked into a second one held down with your finger causes a third penny on the other side of it to fly off even though the second penny never moves.

Equipment

A small clamp, and some pennies and other coins.

Discussion

Hold one penny (A) down with modest pressure from your index finger on a flat surface such as a table top or the OHP. Place a second penny (B) on the surface in direct contact with A. If you flick a third penny (C) directly into A, you will find that B flies off the other side. The collision between C and A created a shock wave that traveled through A and caused B to carry away most of the original momentum. Here are some variations on this demo.

- Let the held coin A be a quarter, while B and C are still pennies. In this case, only part of the shock wave energy and momentum are transmitted to B. The rest of the wave gets reflected back to coin C which recoils backward.
- Replace the single coin A by a row of three coins held in contact by your first two fingers pressed down on the first and last coins. As before, place coin B at one end of the row, and flick coin C into the other end. In this case the shock wave travels from coin to coin, and coin B flies off one end when C hits the other end.
- Hold down a single penny (A) as in the original version, but have two pennies (B1) and (B2) in contact with it. When you flick penny C into the other side of A, you find results that vary enormously from trial to trial. Sometimes B1 carries off most of the momentum, sometimes B2, and sometimes they each take roughly equal shares. The reason for the large variations is that it is impossible to make B1 and B2 exactly come into contact with A. In practice, each is a small distance away, and the relative sizes of these two distances has a drastic effect on the relative transmission of shock waves at the A-B1 and A-B2 boundaries.
- Return to the original set-up with coins A, B, and C. This time, however, rather than holding A down with finger pressure, clamp it to the edge of the table with a small clamp. Now when you propel C into A you will find that B no longer flies off the other side. The pressure from the metal clamp has a drastic effect on the shock wave, whose amplitude through the penny is reduced to a negligible value.

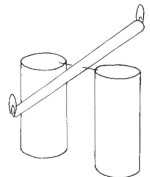

8.18 Burning the candle at both ends

Demonstration
A long candle free to pivot about an axis through its middle and lit at both ends will oscillate about the pivot with a predictable frequency.

Equipment
A long taper or formal candle. If you use a taper candle, its thickness should be around 2 cm at the thick end.

Discussion
To make your candle-oscillator, locate the candle's center of mass by balancing it on your finger, and insert a needle through the center of mass, so as to create an axle. Place the two ends of the axle on the rims of two glasses, with the candle initially in a horizontal orientation. Light both ends of the candle, and after some time the candle will be found to oscillate in a vertical plane. The reason for these oscillations can be traced to the way the rate of burning of the wick depends on the angle of the candle with the horizontal.

For example, suppose the right end of the candle is slightly heavier than the left, and so the right end sinks below the horizontal. The right end flame being vertical will therefore burn the wax at a higher rate than the flame at the left end, and the right end will therefore lose mass faster than the left end. In a short time, this difference in rate of mass loss will result in a torque driving the candle back toward the horizontal. Incidentally, the reason for using thick candles rather than thin ones is that the individual drips of melted wax from a thin candle result in enough of a disturbance to the balance of the candle so that it oscillates in an erratic (nonharmonic) manner.

In contrast, for a thick candle, the motion is much smoother. In fact, as we shall show below, the candle is perpetually experiencing a torque toward its equilibrium orientation, and oscillations about that orientation are the result. For simplicity, let us assume that the candle has a uniform cross section, because this simplifies the analysis. Suppose we designate the mass of the candle as m, and the masses of the right and left "halves" of the candle as m_L and m_R. The center of gravity of each "half" is located approximately a quarter the length of the candle, L, from the center. Therefore, we can write the net torque due to gravity about the center as

$$\tau = \frac{1}{4}gL(m_R - m_L). \tag{8.4}$$

Let us assume that for small angles θ, the rate of mass loss during burning is a linear function of the angle $dm/dt = \alpha - \beta\theta$. The negative sign in the previous equation indicates that for negative values of the angle (where the end dips below the horizontal), the candle burns faster. But, since the right and left ends of the candle make opposite sign angles with the vertical, we have that $dm_R/dt = -dm_L/dt = \alpha - \beta\theta$. Differentiating equation 8.4 with respect to time, we therefore find

$$\frac{d\tau}{dt} = \frac{1}{4}gL\left(\frac{dm_R}{dt} - \frac{dm_L}{dt}\right) = -\frac{1}{2}\beta gL\frac{d\theta}{dt}. \qquad (8.5)$$

Taking the candle to be a uniform rod suspended about a point very close to its center, we have for its moment of inertia $I = \frac{1}{12}mL^2$. Then applying Newton's second law $\tau = Id^2\theta/dt^2$, we find using equation 8.5 that

$$\frac{d\tau}{dt} = I\frac{d^3\theta}{dt^3} = \frac{1}{12}ML^2\frac{d^3\theta}{dt^3} = -\frac{1}{2}\beta gL\frac{d\theta}{dt} \qquad (8.6)$$

so that

$$\frac{d^3\theta}{dt^3} = -\omega^2\frac{d\theta}{dt}, \qquad (8.7)$$

where $\omega = \sqrt{\frac{6\beta g}{mL}}$. The solution to equation 8.7 is an oscillation in angle θ with frequency ω. To understand how the oscillation frequency ω depends on the length of the candle, let us write the candle's mass m in terms of the density ρ and cross-sectional area A, that is, $m = \rho AL$. We therefore find for the oscillation frequency

$$\omega = \frac{1}{L}\sqrt{\frac{6\beta g}{\rho A}}. \qquad (8.8)$$

Notice that equation 8.8 predicts an oscillation frequency that is inversely proportional to the length of the candle, so the oscillations should speed up as the candle length decreases during burning. Alternatively, if you did the demo using two candles, one of which was half the length of the other, you should find that the shorter candle oscillated twice as fast.

Mechanical Oscillations and Waves

One particularly interesting aspect of this demo is the delayed nature of the oscillations, which may not occur until after the ends of the candle have been burning for as much as a minute. The explanation for this behavior lies in the fact that the candle is initially cylindrically symmetric, so that the mass loss initially depends only very weakly on its angle with the horizontal—not enough to bring about oscillations perhaps. However, as it burns, the melting wax carves out hollows on the upper surface of the candle at its two ends.

As a result of these hollows, if one end of the candle tips slightly downward, the flame there consumes wax at a much greater rate than it did originally, and the size of the restoring force that causes the oscillations increases. In effect, the size of the β parameter increases with time, with the result that both the amplitude and frequency of the oscillations also increase over time. This demo was suggested to me in its qualitative form by Randy Elde.[9]

Notes

1. Gren Mason presented this variation of the demo at a local meeting of the British Columbia section of the American Association of Physics Teachers.

2. D. Wagner, T. Walkiewicz, and D. G. Hinan, The American Journal of Physics, **63**, 1014–17 (1995).

3. P. Ceperley, The American Journal of Physics, **60**, 938–42 (1992).

4. It is worth noting that in the rotating wave, the water itself is not rotating, merely the wave disturbance, as you can easily verify by placing bits of cork or styrofoam on the water surface. Thus, you do not want to try to excite the wave by rotating the tank itself.

5. The appearance of the factor $e^{i(m\phi - \omega t)}$ in the wave function explicitly shows that for $m \neq 0$, we are dealing with a rotating wave mode.

6. P. Doherty and D. Rathjen, eds., *The Exploratorium Science Snackbook,* San Francisco, CA, 1991, p. 83-1.

7. Another rare material that shrinks when heated, a zirconium tungstate compound, was reported in the June 1996 Physics Today.

8. G. Biehl, The Physics Teacher, **27**, 172 (1989).

9. Randy Elde in turn credits Mike Lyman who performed the demo at a meeting of the Wisconsin South West Area Physics Sharing Group (SWAPS).

Chapter 9

Electricity and Magnetism

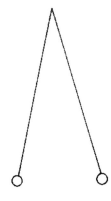

9.1 Coulomb's law

Demonstration
The inverse square distance dependence of the electrostatic force between point charges can be tested using a pair of charged pith balls.

Equipment
Two small pith balls approximately a centimeter in diameter hanging from a pair of one-foot-long nylon threads attached to a common support, and a glass or ebonite rod to charge the balls.

Discussion
The viability of many static electricity demos—including this one—depend greatly on weather conditions. Although two lengths of sticky tape peeled straight off the reel can give a nice *qualitative* demo of electrostatic repulsion even in humid environments, it does not illustrate the inverse square law, as this demo does. Hang the pith balls from nylon threads suspended from a common point, and place as great an electrostatic charge as you can on the balls. Observe that the two strings no longer hang vertically due to the ball's mutual electrostatic repulsion. Measure the horizontal separation between the ball centers, r_1, and also the vertical distance, y_1, from the point of support to the line joining the two balls. Now, sandwich the string between your thumb and index finger, and slide your two fingers down the strings to approximately the halfway point, so that the distance y is halved, that is, $y_2 = y_1/2$. Measure the new separation between the balls, r_2. As we will show, Coulomb's law predicts that the two measured separations should be related according to $r_2 = 1.26r_1$.

You may wish to repeat the demonstration, only this time slide your two fingers down the strings to approximately the 3/4 point, so that $y_3 = y_1/4$, and again measure the separation between balls, r_3, which Coulomb's law now predicts should be given by $r_3 = 1.59r_1$. These predictions, of course, assume point charges, massless strings, and no other sources of electrostatic force. Therefore, the degree of agreement between your measurements and the prediction may be poor if the size of the pith balls is not a small fraction of their separation, if their weight is not large compared to that of the threads, or if other electrostatic forces are present. The remainder of this section will be used to prove the relations given above for the predicted separations.

Consider the three forces acting on each pith ball: gravity (mg), the horizontal electrostatic force (kq^2/r^2), and the tension in the string (T), which makes an angle θ with the vertical. If the pith ball is not accelerating, these three force vectors add to zero, and they form a closed triangle. Thus, we have that

$$\tan \theta = \frac{kq^2/r^2}{mg}.$$

But, from geometry $\tan \theta = \frac{r}{2y}$, so the previous equation can be put into the form

$$\frac{r^3}{y} = \frac{2kq^2}{mg}.$$

If we make several measurements on the balls without changing their charge, the right-hand side of this equation is a constant, so that we have

$$\frac{r_1{}^3}{y_1} = \frac{r_2{}^3}{y_2} = \frac{r_3{}^3}{y_3}.$$

But, since $y_2 = y_1/2$, and $y_3 = y_1/4$, we therefore have $r_2 = 2^{1/3}r_1 = 1.26r_1$, and $r_3 = 4^{1/3}r_1 = 1.59r_1$.

9.2 A 3-4-5 triangle of resistance

Demonstration
Measurement of the resistances of the sides of a 3-4-5 triangle made of high-resistance wire provides an exercise in the series and parallel addition of resistance.

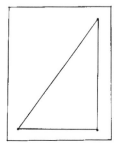

Equipment
A multimeter (digital, if possible), three nails or screws, some high-resistance nichrome wire, and a board or piece of lucite if you wish to do the demonstration on the OHP.

Discussion
Put the nails or screws into the board so that they form the vertices of a 3-4-5 right triangle—perhaps making the actual triangle dimensions 12-16-20 centimeters. Connect nichrome wire between each pair of nails, wrapping the wire securely around each nail, and keeping it taut between nails. What do you expect to find when measuring the resistances along each side of the triangle?

If the sides were not connected, the ratio of the three resistances would equal 3 : 4 : 5, because the resistance of a piece of wire is proportional to its length. However, because the sides are connected, when you measure the resistance along the 3 side you are actually measuring the resistance of side 3 in parallel with the sides 4 and 5 in series. In other words, if we connect the meter across side 3, we would measure a value R_{3meas} given by

$$\frac{1}{R_{3meas}} = \frac{1}{R_3} + \frac{1}{R_4 + R_5} = \frac{C}{3} + \frac{C}{4+5} = \frac{12C}{27},$$

where C is a proportionality constant that represents the length of a one-ohm resistor. Likewise, the measured value of the resistances along sides 4 and 5 can be found from

$$\frac{1}{R_{4meas}} = \frac{1}{R_4} + \frac{1}{R_3 + R_5} = \frac{C}{4} + \frac{C}{3+5} = \frac{12C}{32},$$

and

$$\frac{1}{R_{5meas}} = \frac{1}{R_5} + \frac{1}{R_3 + R_4} = \frac{C}{5} + \frac{C}{3+4} = \frac{12C}{35}.$$

Therefore, we predict that a measurement of the resistances along the three sides of the 3-4-5 triangle should yield resistances having the ratios 27 : 32 : 35. A close agreement with this prediction confirms the procedure for adding resistances in series and parallel.

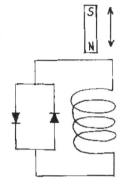

9.3 Induced currents using LEDs

Demonstration
Induced currents can easily be demonstrated by bringing a magnet near a coil to which a pair of LEDs (light-emitting diodes) are connected.

Equipment
A coil of wire having several thousand turns, a bar magnet, and two LEDs, one red and one yellow, such as Radio Shack's 276-041 (red) and 276-021 (yellow).

Discussion
In the usual demo given of induced currents, in which a magnet is moved near a coil of wire, the current is sufficiently small that a sensitive galvanometer is required to reveal its presence. However, as explained in an article by Dan Lottis and Herbert Jaeger, LEDs, which light using far less current than a small incandescent bulb, can do the job quite well.[1] If you connect the red LED and yellow LED in parallel, but with opposite polarity, across the ends of the coil using alligator clips, one will light for one current direction, and the other will light for the opposite current direction. If you furnish the audience with the strength of the magnetic field B, and the fact that LEDs require about a volt to light, it should be possible for them to use Faraday's law, $V = -Nd\phi/dt = -NBA/\Delta t$, to estimate roughly how short the time for the magnet to move away, Δt must be in order for the LED to light.

9.4 A generator and a capacitor

Demonstration
When you connect a hand-crank generator to a capacitor, you will find that it keeps turning in the same direction when you release the crank.

Equipment
A hand-crank generator and a one-farad capacitor, both of which can be obtained from scientific equipment companies.

Discussion

Turn a hand-crank generator, and you will find that it stops almost immediately when you release the crank, because you must supply energy to create the generator's EMF. Now, connect the output wires to the two plates of a one-farad capacitor, and turn the crank a little while. When you release the crank it will keep turning for a while in the same direction. How can this strange behavior be explained?

The generator charges the capacitor, which then discharges after you stop turning the crank, producing a current opposite to the direction of the original current. This current causes the generator to turn, because all generators are also motors, and they rotate when a current is fed into them. The sign conventions for the relation between the current direction and the direction of motion are opposite for motors and generators (right- and left-hand rules), so reversing the current direction keeps the direction of rotation the same when the generator becomes a motor. (You need to be careful not to turn the generator too fast in this demo, because one-farad capacitors typically have a maximum voltage of 5.5 V.)

9.5 Magnetic field of a dipole

Demonstration

The inverse cube variation of the magnetic field of a dipole on distance can be verified using a transparent compass on the OHP.

Equipment

A transparent liquid-filled compass and a small disk magnet.

Discussion

A compass needle in the presence of ambient magnetic fields, B_E (mainly due to the earth and any nearby iron), will align itself to point along a direction we shall call north. Now, suppose a compass is also in the presence of another magnetic field B due to a nearby magnet which points toward the east at the location of the compass. Due to the two perpendicular fields, the direction of the net magnetic field (the direction along which the compass will point), makes an angle whose tangent with respect to the north is given by $\tan \theta = B/B_E$, based on simple vector addition. Thus, we can infer the variation of B with distance from the magnet, based on observations of $\tan \theta$. In particular, for a dipole field for which B drops off as the inverse *cube* of the distance from the magnet, we predict that when the distance from the magnet to the compass is doubled, the tangent of the compass needle deflection from the north drops to one-eighth its original value.

In order to check this prediction, place the compass on the OHP at a location so there is at least 25 centimeters on the glass along a direction either to the east or west of the compass. Orient the compass so that the needle points to the N. Now, place the disk magnet along an east-west line passing through the center of the compass, with the axis of the disk facing the compass. Vary the distance of the magnet from the compass until the compass needle deflection from the north is 60 degrees. Now, move the magnet twice as far away from the compass along an east-west line, and observe the deflection angle θ from the north. If the magnet has a dipole magnetic field, you should find that the tangent of θ is one-eighth the tangent of 60 degrees, or $\theta = 12.2$ degrees. This demo was suggested to me by Bruce Sherwood, based on another version by Ron Edge.[2]

9.6 Pushing disk magnets on the OHP

Demonstration
The motion of small disk magnets on the overhead projector has many interesting aspects.

Equipment
A collection of ceramic disk magnets (of diameter about a centimeter), which can be purchased from a scientific equipment company or possibly a toy or hardware store.

Discussion
Observation 1. Place two disk magnets face down on the OHP with like poles facing down, so that the magnets repel each other. Move one toward the other slowly, and notice how the other magnet retreats—always keeping roughly the same distance. Actually,

a more careful observation indicates that the retreating magnet moves in a series of steps rather than at a continuous speed. Why does it move in steps? The steps are a consequence of the difference between the value of the static and sliding friction coefficients, μ_s and μ_k. A magnet brought near a second initially stationary magnet will not cause the second magnet to move until the repulsive force is sufficient to overcome static friction, at which point the second magnet will accelerate away with an initial acceleration $a = (\mu_s - \mu_k)g$. The retreating magnet will soon come to rest because the initial acceleration becomes a deceleration as the repulsion weakens with distance.

Observation 2. Place three disk magnets face down on the OHP, all with the same polarity, so that they all repel one another. Place your two index fingers on two of them, and move them toward the third one, which is observed to retreat. Keep the two "driving" magnets the same distances from the retreating magnet. Notice how the direction in which the retreating magnet moves lies along the bisector of the angle formed by the lines from each driving magnet to the retreating magnet—see figure—as required by the vector addition of the two repulsive forces, which have equal magnitude if the distances are the same.

Observation 3. Place three disk magnets face down on the OHP, again with all having the same polarity, so that they all repel one another. Place them in a straight line, and push an end one (the "driver") with your index finger toward the middle one. As the middle one starts to move it will drive the front one forward, so that the three remain in a row as you drive the two forward. You will probably find that to keep them moving in the same direction you need to continually adjust the location of the driving magnet to compensate for small misalignments that get magnified as a result of the motion. (In other words, a small error in the direction of push on the middle magnet causes a larger error in the direction the middle magnet in turn pushes on the front one.)

Observation 4. Repeat observation 3, but now observe the spacing between the three magnets as the driving magnet is pushed forward. You should find that the average spacing between the driving magnet and the middle one is slightly less than that between the middle magnet and the front one. The difference in spacing is a result of the different repulsive forces between the magnets. When front and middle magnets are both just at the point of moving, the driving magnet exerts approximately twice the force on the middle magnet as the middle one exerts on the front one (because in pushing on the middle magnet the driving magnet is also, in effect, pushing on the front one). As a result, the driving magnet needs to get a bit closer to the middle magnet for the force to double—not much closer, because the force for a pair of dipoles varies approximately as the inverse fourth power of the center-to-center separation. Thus, a distance ratio of $2^{1/4} = 1.19$ to 1 will give a force ratio of 1 to 2. (When I measured the distance between the magnets *while* pushing the driver magnet, the distance ratio was in fact found to be 1.2, in excellent agreement with the predicted ratio.)[3]

Observation 5. Place *four* disk magnets face down on the OHP, again with all having the same polarity, so that they all repel one another. Place them in a straight line, and push an end one (the "driver) with your index finger toward the one ahead of it. Try as you will, you very likely will not be able to push the three magnets ahead of the driver moving in a row, as you were able to do in observation 3. The magnification of misalignments referred to in observation 3 is now much more serious with one extra magnet in the row. Unless you are extremely skillful, you will find that moving the driver magnet from side to side to restore the alignment of the magnets ahead no longer works.

9.7 A magnetic oscillator

Demonstration
The period of a simple magnetic oscillator can be measured and compared with theory.

Equipment
A metronome, several index cards, and four small (1 centimeter diameter) disk magnets that can be purchased from scientific equipment companies and toy, nature, or hardware stores.

Discussion
You can make the magnetic oscillator shown in the figure from two index cards and four disk magnets.

Fold two 3-by-5-inch index cards lengthwise, creating two double-thickness cards having dimensions 1.5 by 5 inches. For each card, tape two of the magnets to two of the adjacent corners which are spaced 1.5 inches apart. The polarities of the magnets should be such that when the cards are taped together at the ends without the magnets, a mutual repulsion between the magnets keeps the cards apart at their magnet ends. To create oscillations of the device, you need only push down on the top card, and then release it. Assuming that you aligned the magnets with care, and didn't use too much tape in taping the card ends together, you should find long-lasting oscillations about some equilibrium angle—see figure. As we will show later, the predicted period of these oscillations is

$$T = \frac{T_0}{2}\sqrt{\theta_0} \qquad (9.1)$$

where the angle θ_0 is the equilibrium angle in radians, and T_0 is the period of the cards when they swing as a physical pendulum. In order to test equation 9.1, you need to measure the two periods T and T_0, and the angle θ_0. To make the physical pendulum out of the device, simply hang the two cards (magnet side down) draped over a straightened paper clip held horizontally. You can then simply count the number of pendulum swings in one minute to measure T_0.

Measuring the period T of the magnetic oscillator is a bit trickier, because these oscillations occur at a frequency too high to count individually. However, if you set a metronome to match the oscillations, you can measure the period based on the observed matching metronome frequency. To test equation 9.1, you also need to measure the equilibrium angle of the device θ_0. For small values, the angle can be found from $\theta_0 = d/L$, where d is the vertical (center-to-center) separation between the magnets, and L is the length of the cards.

Here are some results from one such experiment. The equilibrium angle θ_0 found from d/L was 1/8 radians, and the period of pendulum oscillations was measured as 1.4 seconds, based on a count of the number of full pendulum swings in one minute. Using equation 9.1, we therefore predict a period of 0.25 seconds for the magnetic oscillator. When the oscil-

lator period was measured by metronome, the result
was found to be 0.29 seconds—in reasonable agreement with the predicted value—particularly in light of
the numerous approximations made in deriving equation 9.1. The remainder of this discussion will be devoted to a somewhat lengthy derivation of that equation.

Let us make the following assumptions:

• The angle of the oscillator is small, so that the vertical separation between magnets can be expressed
as $y = L\theta$.
• The oscillations have small amplitude, so that $\theta = \theta_0 + \epsilon$, where $\epsilon << \theta_0$.
• The force between the magnets is a dipole-dipole
interaction, so that $F = k/y^4$.

In view of these assumptions, the force between the
magnets can be expressed as

$$F = \frac{k}{y^4} = \frac{k}{(L\theta)^4} = \frac{k}{(L\theta_0)^4}\left(1 + \frac{\epsilon}{\theta_0}\right)^{-4}, \qquad (9.2)$$

which to a first order expansion in ϵ/θ_0 yields

$$F = \frac{k}{(L\theta_0)^4}\left(1 - 4\frac{\epsilon}{\theta_0}\right). \qquad (9.3)$$

Applying Newton's second law $\Sigma\tau = I\frac{d^2\theta}{dt^2}$ to the magnet oscillator yields

$$-mg\ell + FL = I\frac{d^2\theta}{dt^2}, \qquad (9.4)$$

where m and I are the mass and moment of inertia
of the top card of the oscillator, L is the length of the
card, and ℓ is the distance from the center of mass to
the pivot at the taped end. Substituting equation 9.3
into 9.4, and using $d^2\theta/dt^2 = d^2\epsilon/dt^2$, we obtain

$$-mg\ell + \frac{kL}{(L\theta_0)^4}\left(1 - 4\frac{\epsilon}{\theta_0}\right) = I\frac{d^2\epsilon}{dt^2}. \qquad (9.5)$$

Applying equation 9.5 to the special case where the
oscillator is at rest ($\epsilon = d^2\epsilon/dt^2 = 0$), yields the result

$$mg\ell = \frac{kL}{(L\theta_0)^4}. \qquad (9.6)$$

Finally, using equation 9.6, we can rewrite equation 9.5 as

$$-\frac{4mg\ell}{I\theta_0} \equiv -\omega^2 = \frac{d^2\epsilon}{dt^2}. \tag{9.7}$$

The solution of equation 9.7 is simple harmonic oscillations with frequency ω and period $T = 2\pi/\omega$. Thus, based on the definition of ω in equation 9.7, the period T can be written as

$$T = 2\pi/\omega = \frac{2\pi}{\sqrt{\frac{4mg\ell}{I\theta_0}}} = \frac{T_0}{2}\sqrt{\theta_0}, \tag{9.8}$$

where $T_0 = 2\pi\sqrt{I/mg\ell}$ is the period of oscillations of the cards as a physical pendulum.

Notes

1. D. Lottis and H. Jaeger, The Physics Teacher, **34**, 144-46 (1996).

2. Sherwood uses the demo as a "desktop experiment." The other version of the demo appeared in R. D. Edge's book, *String and Sticky Tape Experiments,* American Association of Physics Teachers, College Park, MD (1981), p. 11.05.

3. The analysis described ignores the force acting between the driver magnet and the front one, which is roughly only 1/16 as much as that between the driver and the middle magnet. The closeness between the observed and predicted spacing ratio shows this to be a good approximation.

Chapter 10

Optics

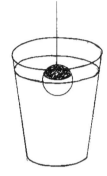

10.1 Soot-covered ball underwater

Demonstration
A soot-covered metal ball suspended underwater appears silvered.

Equipment
A candle, a metal ball on a string, and a glass of water.

Discussion
Cover a metal ball at the end of a string with black soot by holding a candle under it. Once it is soot-covered, submerge the ball under water, and it will appear silvered, rather than black. The soot covering the ball has many small nooks and crannies that trap a layer of air next to it. Nearly all the light incident on this smooth air layer is reflected, causing it to have a silvered appearance.

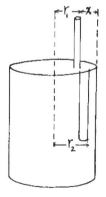

10.2 Finding *n* using a pencil in a glass of water

Demonstration
The apparent break in a cylindrical object placed off-center in a water-filled container can be used to measure the index of refraction of the water (or other liquid).

Equipment
A plastic ruler, a cylindrical glass (or jar) of water, and a cylinder (such as a *round* pencil), whose diameter is no more than about a tenth that of the glass. The larger the glass (or jar) diameter, the greater the precision. A narrow popsicle stick or other flat object works just as well instead of a cylinder, but we shall refer to the object as a pencil below.

Discussion

Hold the pencil at the center of the water-filled cylinder with half its length submerged, and view the pencil with your head at the same horizontal level as the waterline. It can easily be shown that the magnification of the submerged portion of the pencil located at the center (or in a plane through the center) is n, the index of refraction of water. As a result, the submerged portion of the pencil will be noticeably wider than the portion above the waterline. Now, move the pencil away from the center at right angles to your line of sight, and you will notice an apparent break in the pencil at the waterline.

When the pencil is located at some critical distance x from the edge of the glass, the break will be sufficient to just appear to shift the underwater image of the pencil by one pencil width from the portion above the waterline—see figure. If it were moved any closer toward the edge of the glass, the upper and lower portions of the pencil would appear completely severed—try it. (This effect can only be observed if the pencil width is not greater than about a tenth the diameter of the glass.) Record the distance x of the right edge of the pencil from the edge of the glass at the critical distance. In order that your measurement be reliable, it is essential that the center of the pencil lie along a *diameter* perpendicular to your line of sight. (You might want to lay a ruler on top of the glass to help position the pencil properly during its motion.)

As shown below, the index of refraction of the water can be found from the following equation

$$n = \frac{D/2 - x}{D/2 - x - d}, \tag{10.1}$$

where D and d are the diameters of the glass and pencil, respectively. When I did the experiment using an 8-cm-wide glass and a 0.7-cm-wide pencil, I found that $x = 1.0$ cm. Using these values, equation 10.1 gives $n = 1.3$, which is as close to the standard value $n = 1.33$ as measurement uncertainty would allow here. In order to prove the equation for n, note that (1) the width of the magnified underwater portion of the pencil is nd, and (2) $r_2 = nr_1$ (see figure). Therefore, $nd = r_2 - r_1 = nr_1 - r_2$. If we use $r_1 = D/2 - x$

in the preceding equation, and solve for *n*, we obtain equation 10.1.

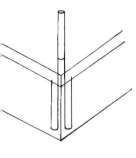

10.3 Seeing double in a square container

Demonstration
A pencil placed in a water-filled square container will be seen as double when the container is viewed along a direction joining two opposite diagonals of the square container.

Equipment
A square water-filled container whose corners are not rounded, and a pencil. (The container could be rectangular instead of square, such as a small aquarium or transparent plastic shoe box.)

Discussion
Place a pencil vertically in the water, and view its underwater portion from some distance away, while looking directly at one corner of the container. You need to orient the container and the pencil so that the two sides nearest you make 45 degrees with your line of sight, and the pencil is directly along your line of sight. If the container is square rather than rectangular, then the pencil and the far opposite corner of the container are both along your line of sight.

You should see two images of the pencil—one through each of the two sides of the container. The separation of the two pencil images depends on both the index of refraction of the water and the distance of the pencil from the corner. This observed separation increases, the greater the index of refraction of the liquid, and the greater the pencil's distance from the corner—as you can easily verify by moving the pencil away along your line of sight.

Using Snell's law, it can be shown that the index of refraction of the water can be expressed as

$$\frac{1}{n} = \sqrt{2}\sin\left[\frac{\pi}{4} - \tan^{-1}\left(\frac{1}{\frac{d}{D} - 1}\right)\right] \tag{10.2}$$

where $\tan^{-1}$ is the arctangent function, *D* is the pencil diameter, and *d* is the distance the *center* of the pencil is behind the corner in order for the two pencil images to be one pencil width apart. Be sure that you measure the distance d from the center of the pencil to the *outside* of the corner—in effect, treating the glass as though it were part of the water. Using equation 10.2, we can find values of n computed for $d/D = 4.0$, 4.5, 5.0, 5.5, and 6.0. The corresponding n values are $n = 1.58$, 1.46, 1.37, 1.32, and 1.27. Thus, given the accepted value $n = 1.33$ for water, you should find that the center of the pencil needs to be placed a distance just under 5.5 pencil diameters behind the corner to achieve the desired separation in the two images. (The exact value is 5.34 pencil diameters.)

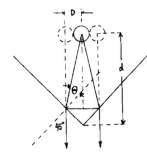

To test this prediction, just move the pencil until the gap between the two underwater images is just matched by the portion of the pencil above the water line, and convert your measured value of d/D to an index n either by substitution in equation 10.2, or by interpolating between the values listed above for different n values. You might try using two values of d/D corresponding to your estimate of the measurement uncertainty, and see if the two values bracket the accepted value $n = 1.33$. (Which measured quantity D or d would you expect to be the main contributor to uncertainty in the ratio d/D?) Note that equation 10.2 holds for arbitrary pencil radii, so you may get a more accurate measurement using a vertical cylinder having a radius somewhat larger than a pencil. The larger the diameter of the cylinder, however, the larger the container needs to be in order to place the cylinder so that its center is a distance 5.34 diameters away from the corner.

In order to derive equation 10.2, suppose the two images of the pencil are in fact one pencil width apart. The light rays from your distant eye projected back to these images make an angle of 45 degrees with the two side walls of the container. Using Snell's law, the angle of refraction in the water for a 45-degree ($\frac{\pi}{4}$ radian) angle of incidence is given by $\sin \theta = \frac{1}{n} \sin \frac{\pi}{4} = \frac{1}{n\sqrt{2}}$, so that $\frac{1}{n} = \sqrt{2} \sin \theta$. Applying some simple geometry to the figure, the distance from the corner of the container to the center of the pencil can be expressed as $d = D(\cot(\frac{\pi}{4} - \theta) - 1)$, where D is the pencil's diameter. If this equation is solved for $\frac{1}{n} = \sqrt{2} \sin \theta$, equation 10.2 is the result.

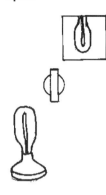

10.4 Partially blocking a converging lens

Demonstration
If you partially block a converging lens, no part of the image is obscured; rather, the image formed by the lens on a screen merely gets dimmer.

Equipment
A large converging lens, such as a magnifying glass, an unfrosted light bulb (to serve as the object), and a screen.

Discussion
Form an image of the light bulb filament on a screen. When asked to predict what happens when half the lens is covered, many people believe that half the image will disappear, rather than the whole image merely getting fainter, and they are quite surprised by the result. To do the demonstration, you might try using a blocking piece of paper about half the width of the lens. Place the paper across the center of the lens, and move it slowly from the lens to the screen behind the lens, where it does in fact block a portion of the image that appears on the paper rather than the screen. Now slowly move the paper back toward the lens, and notice how the two edges of the paper's shadow, initially quite sharp, go out of focus and re-join to form the complete image. This demonstration is based on an article by Jay Huebner.[1]

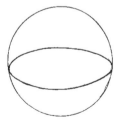

10.5 A water lens

Demonstration
A plano convex water lens can easily be made using a transparent plastic globe.

Equipment
Two joined transparent plastic hemispheres having a diameter of about 10 centimeters. Such plastic globes may be purchased at crafts stores.

Discussion
If you add a few centimeters depth of water to one of the plastic hemispheres, and then join them at their seam (you may also want to tape them together), the

device acts like a water lens of radius $R = 5$ centimeters, and focal length $f = R/(n-1) = 5/0.33 = 15$ cm. To verify the focal length, simply place the globe on a ring stand or else some transparent surface, such as a sheet of lucite, below which you can place a piece of paper. Hold a lamp with an unfrosted filament about a meter above the lens. The filament image should appear on a piece of paper placed below the lens. Measure the distance from the bottom of the globe to the image, and see how well it agrees with the formula for the focal length. (This water lens is thick enough that you should probably add half its thickness to the measured focal length in order to get a more accurate comparison with theory.)

It might also be interesting to try using a layer of oil floating on a layer of water. This combination should provide a lens having two focal lengths, because in the central region the water makes contact with the curved plastic surface, while in the outer region the oil makes contact. In effect, we have two separate concentric lenses with different focal lengths. The focal length of the oil part of the lens may be found using the above formula, replacing $n = 1.33$ by the index of refraction of oil. You should be able to measure the two focal lengths using the procedure mentioned in the preceding paragraph.

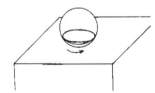

10.6 A rotating water lens

Demonstration
A rotating container of water on the OHP reveals the change in the curvature of the water surface through the change in the focal length of the water lens.

Equipment
Two joined transparent plastic hemispheres having a diameter of 10 centimeters. Such plastic globes may be purchased at crafts stores.

Discussion
After adding a few centimeters depth of water to one of the hemispheres, and joining them together, as described in the previous demonstration, place the water-filled globe on the overhead projector. You should observe a bright circular region in the center of the image on the screen, which represents an

out-of-focus image of the projector filament, as you can easily verify by bringing it into focus.

Now, go back to the original focus, and give the sphere a hard spin. Once the water stops sloshing, and is spinning smoothly, you should notice that the size of the bright circular region is larger than when the water was at rest. This effect is due to the curvature of the surface of the spinning water, which increases the focal length of the water lens. (To see why the focal length increases, consider what the focal length would be if the water is spinning fast enough for its surface to have the same radius of curvature as the globe itself.)

This setup is a more feasible way of demonstrating the curvature of the water in a spinning container than that suggested in a demo in *Turning the World Inside Out*.[2] However, unlike that earlier demonstration which used a phonograph turntable to provide a known rotational speed, here we have no way of correlating the shape of the water surface (or the focal length of the lens), with the rotational speed, which is not measured.

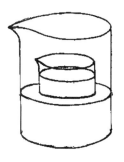

10.7 Invisibility using equal indices of refraction

Demonstration
A glass beaker becomes invisible when immersed in a liquid having the same index of refraction.

Equipment
Some baby oil, and two pyrex glass beakers (one large and one small).

Discussion
Transparent materials are not invisible because light refracts when encountering them. Of course if the material is surrounded on both sides by another transparent material of the same index of refraction, no refraction occurs, and the material will be literally invisible. One way of doing this classic demonstration of invisibility uses two glass beakers and some baby oil, as described in *The Dick and Rae Physics Demo Notebook*.[3]

Try to find a small beaker that has some writing on it. Put the small beaker inside the large one, and pour some baby oil into the small beaker. Now pour some baby oil into the large beaker, so that the lower part of the glass of the small beaker is surrounded on both sides by baby oil. This portion of the small beaker becomes invisible as the oil level rises.

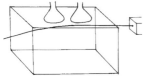

10.8 Mirage formation

Demonstration
Mirages and their cause—the deflection of light rays in a medium of variable index of refraction—can be observed using a small aquarium.

Equipment
A small aquarium, two infrared heat lamps, and a light source producing a roughly parallel light beam, such as a focusable flashlight or laser.

Discussion
Mirages are a common sight—particularly on hot summer days when looking ahead some distance on a long flat highway. The appearance of what looks like water in the distance is due to the refraction of light rays from the sky that encounter the layer of air that is much warmer near the pavement. This air is less dense and has a lower index of refraction than the air higher up. As a result, a light ray from the sky headed downward at a grazing angle to the pavement can be deflected upward to your eye—essentially, you see light from the sky even though your line of sight is toward the pavement.

A nice technique for demonstrating the curvature of a light beam in a medium having a variable index of refraction was suggested in an article by P. R. Barker, P.R.M. Crofts, and M. Gal.[4] The term "superior mirage" refers to the fact that the object appears higher than its actual position, instead of lower. The mirages you see on highways are "inferior" because you see the sky on the roadway—in what appears to be pools of water. Just as inferior mirages require lower layers of the medium to have the lower index of refraction (caused by warming), a superior mirage requires that upper layers of the medium have the lower index.

Barker, Crofts, and Gal's suggested method for producing the needed refractive index variation with height is to fill a small aquarium with water and to place two infrared heat lamps in a stand, so that the lamps are only about a centimeter from the water surface. Surprisingly, the warmth does not penetrate very far into the water, so a sizable temperature gradient and an index of refraction gradient are created in the water.[5]

Now, position a laser or focusable flashlight right next to the tanks, so that it creates a horizontal light beam just below the water surface (before you turned the lamps on). The spot of the beam should be visible on a wall or screen a few meters behind the tank. You should find that the position of the spot noticeably drops on the screen once the lamps are turned on, showing that the curvature of the light beam increases as the temperature gradient becomes greater. If you are using the flashlight beam, you should notice that the shape of the spot on the wall distorts while it drops.

In order to actually see a mirage using this apparatus you need to view an object (such as a frosted light bulb) through the tank when it is placed about a meter from the tank at approximately the same horizontal level as the water. Look through the tank just below the water level, and you should see distorted images of the light bulb.

10.9 An oil spot light meter

Demonstration
An oil spot on a piece of paper can be used as a light-meter to test the inverse square law of illumination.

Equipment
Three unfrosted bulbs, one of which has different wattage from the other two.

Discussion
Place a spot of oil on a piece of paper, and notice how the spot appears dark when the paper is illuminated from the same side as you view it, while it appears light when illuminated from behind. This appearance of the spot is a result of it being a poorer reflector and better transmitter of light than the rest of the

paper. Obviously, the spot will appear the same as the rest of the paper only when the illumination from each side is equal. To verify this prediction place the paper in a vertical frame and place the two equally bright bulbs an equal distance away from the paper on each side, and see if the spot is invisible. It might also be worthwhile to see how different the two distances need to be before you can just distinguish the oil spot from the paper.

Now, use the two bulbs having different wattage. What would we predict for the relative distances of each bulb on each side of the paper in order not to see the oil spot? Let us assume that the brightness of each bulb is directly proportional to its wattage, W, and also that the light intensity I at some distance from the bulb obeys the inverse square law. In other words, given that the area of the sphere over which the light is spread is simply $4\pi r^2$, we can express the light intensity of bulb 1 at some distance r_1 as $I_1 = W_1/4\pi r_1^2$. A similar equation gives the light intensity of bulb 2 at a distance r_2, namely, $I_2 = W_2/4\pi r_2^2$. Therefore, if the two light intensities are equal we must have

$$W_1/r_1^2 = W_2/r_2^2,$$

or, equivalently,

$$r_1/r_2 = \sqrt{W_1/W_2}.$$

Thus, we can use this last equation to predict the ratio of the distances you need to place unequal wattage bulbs on each side of the paper so as not to see the oil spot (equal light intensity on each side).

In doing the experiment try three cases—all having the predicted ratio of distances for the given W_1/W_2 ratio. In other words, if the two wattages are in a 1 : 4 ratio, and the predicted distance ratio is therefore 1 : 2, you might try cases for which $(r_1, r_2) = (0.25, 0.5)$, $(0.5, 1.0)$, and $(1.0, 2.0)$ meters. We would predict that in all three cases, the oil spot should be invisible (due to equal light intensity on both sides). However, you are likely to find some departures from this prediction based on the inverse square law. If departures from the prediction are found, can you explain why one or another of the three cases might be expected to give better agreement with the prediction? There are actually many sources of possible disagreement with the inverse square law, some of which occur at small distances, some at large distances, and some at all distances. Which of the following three circumstances do you think would mainly give a disagreement at small, large, and all distances: (*a*) finite source size, (*b*) other sources of illumination in the room, and (*c*) a bulb brightness not being proportional to the wattage?

10.10 Pinhole imaging

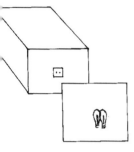

Demonstration
The fuzziness and brightness of images through a pinhole vary in a predictable way with pinhole size.

Equipment
An unfrosted bulb, a cardboard carton about 0.5 meters on a side, and pieces of opaque cardboard in which to make the pinholes.

Discussion
The "quick and dirty way" to do pinhole imaging would be to hold an opaque piece of cardboard (with one or more small holes in it) between a bright source, such as an unfrosted light bulb, and a screen in a darkened room. For images having somewhat better contrast, you may wish to make the "pinhole imaging system," described below. Place the bulb under the cardboard carton with its closed bottom facing upward, so that the light from the bulb would be completely trapped inside the carton. Now, cut a square hole a few centimeters on a side in the center of one side of the carton, so the light from the bulb can get out of the closed box. Make pinholes having a range of sizes—*two* holes in each piece of opaque cardboard which will be taped over the hole in the box.

By using a pair of holes in each piece of cardboard, you will be able to see how the appearance of side-by-side images from two pinholes depends on pinhole size without changing the cardboard. The pair of holes in each piece of cardboard should have diameters in an increasing sequence between about

1 and 8 mm. For example, if you used four pieces of cardboard, the first one might have holes of diameters around 1 and 2 mm, the second one around 2 and 4 mm, the third 3 and 6 mm, and the fourth one 4 and 8 mm. Keep the holes in each piece of cardboard at least a centimeter apart. The image quality should not depend on whether the holes are round.

In order to observe the pinhole images, tape a given piece of cardboard with its pair of pinholes over the square hole in the box, and observe the image of the two pinholes on a screen placed near the box. You could use a piece of paper as a screen, but you may wish to use a piece of ground glass if the screen is to be viewed in rear projection mode. Try to observe what is the largest pinhole that produces an image without serious blurring. Smaller pinhole diameters result in clearer but dimmer images, with the optimum tradeoff between blurriness and brightness ultimately being a function of the taste and eyesight of the demonstrator.

In addition to observing how the image clarity depends on the pinhole size, it would also be useful to show what happens when you vary the distance of the light filament to the pinhole for a given size pinhole. In this case, you should find that (unlike the case of images produced with a lens), the clarity of the image is very insensitive to the distance from the object to the pinhole—in other words, a pinhole camera has a near-infinite depth of field.

Here is one additional demo you can do with the pinhole imaging system. Make a piece of cardboard having many pinholes in it, and place it over the hole in the box, thereby producing many pinhole images of the light bulb filament on the screen. If the screen is placed the proper distance from the pinholes, all the images will coalesce into one when a converging lens is placed over the pinholes.

10.11 The negative pinhole image

Demonstration
Using a small object suspended from a string, you can observe negative images of a bright object on a screen.

169

Optics

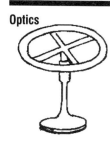

Equipment

A circular fluorescent light (approximately 25 cm in diameter) to serve as the bright object, and two alternative objects to serve as the negative pinhole: a short 1/4-inch (0.5 cm)-diameter screw and a 0.5-cm-diameter sphere—possibly a ball of clay hung by a thread. (Exact dimensions are unimportant.)

Discussion

Because of the pinhole camera, most people are probably familiar with the idea of an image formed by a pinhole placed in front of a screen, but the ability of a "negative pinhole" to form an image is far less well known. A negative pinhole is the exact optical complement of a pinhole—instead of blocking all the light except that headed toward the pinhole, the negative pinhole blocks *only* the light heading toward it—essentially forming a shadow. To form an image of a bright object using a negative pinhole, you need only hang a small object on a string between the bright object and a screen. Although pinholes are often round, their particular shape makes little difference in the appearance of the image—which you can easily verify by alternately using a small sphere or a short screw as the negative pinhole.

You can show by ray tracing that negative pinholes must form complementary images of an object—dark images of bright objects. Essentially, for all points on the negative image, some part of the bright object is completely blocked by the pinhole. Negative images are harder to see than positive ones, however, because of the overall illumination of the screen by the unblocked portion of the bright source. For maximum image visibility, you should use a circular fluorescent bulb, rather than an unfrosted incandescent bulb. (The circular fluorescent bulb also has a very distinctive shape, making it easier to recognize as the image of the object, rather than simply the shadow of the negative pinhole.)

Try varying both the object and image distances to see what combination gives the best negative image on a piece of paper or screen, using the ball and then the screw as the negative pinhole. It may be particularly effective to rotate the light source making the plane of the circular fixture not horizontal, so that its projected negative image on the screen will be an ellipse, rather than a straight line. For best viewing, be sure that all other lights are turned off. More information on the negative pinhole can be found elsewhere.[6, 7]

Notes

1. J. Huebner, The Physics Teacher, January 1989, p. 43.

2. R. Ehrlich, *Turning the World Inside Out*, p. 66.

3. D. R. Carpenter Jr. and R. B. Minnix, *The Dick and Rae Physics Demo Notebook*, p. O-215.

4. P. R. Barker, P.R.M., Crofts, and M. Gal, The American Journal of Physics, **57**, 953–54 (1989).

5. The heat doesn't penetrate very far into the water because when the top layer is warmer, heat can only reach the lower layers through conduction, not via the much more efficient mechanism of convection.

6. D. Bissonnette, P. Rochon, and P. Somer, The Physics Teacher, April 1991.

7. J. E. Stewart, The Physics Teacher, April 1991.

Chapter 11

Interference and Diffraction

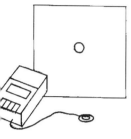

11.1 Speaker in a board

Demonstration
The sound level from a small speaker increases dramatically when it is placed in a hole in a board.

Equipment
A plywood board or piece of cardboard about a meter square, a small speaker (about 3 cm diameter), and a cassette tape recorder.

Discussion
Plug the jack from a tape recorder into the small speaker, and you will find the sound to be very weak. A bare speaker (not placed in a cabinet) produces a very low sound intensity, because as the moving part of the speaker vibrates back and forth, it generates two pressure waves—one toward the listener and a second away from the listener 180 degrees out of phase with the first. If the speaker is bare, there is a partial destructive interference of these two sounds, because the sound from the back of the speaker diffracts (bends) around it and combines with the first sound. The effect is particularly pronounced for small speakers, because the extent of diffraction increases the smaller the size of the speaker. Now, if you place the speaker in a matching size hole in the piece of cardboard or plywood board, the sound level heard by a listener in front of the speaker is dramatically louder, because the waves from the back of the speaker can no longer diffract around the front as well, and hence less destructive interference occurs. This demonstration is from *The Dick and Rae Physics Demo Notebook*.[1]

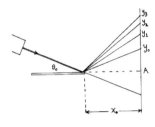

11.2 Measuring the wavelength of light by ruler

Demonstration
A steel ruler illuminated by a laser produces a diffraction pattern that allows the wavelength of light to be measured.

Equipment
A steel machinist's ruler with engraved rulings and a laser or laser pointer. You might want to try several rulers, one with a finer scale than the other.

Discussion
In a classic paper, A. L. Schawlow explained how the wavelength of light can be measured using a steel ruler illuminated by a laser.[2] The ruler needs to be positioned nearly horizontally in front of the laser, so that if the laser is tilted downward slightly its beam makes a small angle with the ruler, and grazes its last few centimeters. The laser beam needs to make a much more grazing angle with the ruler—that is, smaller θ_0—than shown in the figure. The laser needs to be aimed to illuminate the finest gradations on the ruler. These equally spaced ruler gradations effectively make the ruler a diffraction grating of the reflection type, and you should be able to see several orders of diffraction as spots on a distant screen.

In addition, if you point the laser so that some of the beam passes over the end or side of the ruler, you will be able to see on the screen both the reflected beams for various orders, as well as the direct beam heading downward at an angle $-\theta_0$ below the horizontal. For a reflection grating, the zeroth order beam is specularly reflected (as with a mirror), so it leaves the ruler heading upward at an angle $+\theta_0$. Consequently, the plane of the ruler intersects the screen at an identifiable point A midway between the direct laser beam and the specularly reflected beam that hits the screen at point y_0.

In general, the nth order reflected beam satisfies the condition that light reflected off the ruler in a particular direction travels n fewer (or greater) wavelengths from the spaces between two adjacent markings. Using simple geometry, it can be shown that the

preceding condition translates into

$$\lambda = d(\cos \theta_{n+1} - \cos \theta_n), \tag{11.1}$$

where d is the spacing between adjacent ruler markings, and θ_n is the angle the nth order beam makes with the plane of the ruler. Since θ_n and θ_{n+1} are small angles we may use the approximation $\cos \theta_n \approx 1 - \frac{1}{2}\theta_n^2 \approx 1 - \frac{y_n^2}{2x_0}$, where y_n is the position of the nth order on the screen relative to point A, and x_0 is the distance from the ruler to the screen—see figure. Using this approximation in equation 11.1 gives

$$\lambda = \frac{d}{2}\left(\frac{y_{n+1}^2 - y_n^2}{x_0^2}\right). \tag{11.2}$$

You will find that the spacing between adjacent orders on the wall is not uniform, because a uniform spacing of $y_{n+1}^2 - y_n^2$, predicted by equation 11.2 is not the same as a uniform spacing for $y_{n+1} - y_n$. Using equation 11.2 with the measured location y_n of each order, you could see how consistent a value you obtain for the wavelength calculated from each pair of adjacent orders. With a little care, differences from an average value should be in the range of a few percent. Just be sure to measure each y_n distance from point A.

11.3 Colorful soap bubbles

Demonstration
Long-lasting soap bubbles illuminated by white light serve to illustrate the principles of thin film interference.

Equipment
An Erlenmeyer filter flask with a side arm (available from chemical supply companies), Tygon tubing to fit on the side arm, a small spatula, a slide projector, and a soap bubble solution made from one part glycerine, two parts liquid soap, and seven parts distilled water. If you cannot easily get an Erlenmeyer filter flask, you could improvise using a small plastic soda bottle with a hole in its side, into which you have inserted a straw. (Seal the space between the straw and the bottle with plastic cement or modeling clay.)

Discussion

The use of soap bubbles to illustrate thin film interference has been described by numerous authors including this one in *Turning the World Inside Out*.[3] Two problems with soap bubbles for large group demonstrations are their impermanence and the difficulty of viewing them in a large class setting.[4] An article by F. Behroozi and D. W. Olson, on which this demo is based, describes solutions to both of these problems.[5]

Here is their recipe for long-lasting spherical soap bubbles. Partially fill a filter flask with warm water. Dip a spatula in a soap bubble solution, and draw it across the mouth of the flask, creating a thin film across the opening. Gently blow through the Tygon tubing attached to the side arm of the flask, and you will produce a soap bubble perched on the neck of the flask. You can keep the bubble from shrinking by crimping the tubing. Apparently the moisture inside the bubble from the water in the flask is largely responsible for the long life of the bubble.

If the bubble is illuminated by the beam from a slide projector in a darkened room, two intense colorful spots will be seen on the bubble. These spots are images of the slide projector light bulb from the front and back surfaces of the bubble, which act like convex and concave mirrors, respectively. The color of these images will be seen to vary in a systematic way with viewing angle. Spectators should be encouraged to move their heads from side to side to see a perceptible color change. Also, you should systematically vary the angle of illumination of the bubble, so that spectators can observe large variations in color without changing their viewing angle.

The dependence of image color on viewing angle is a result of the constructive interference between light reflected off the two surfaces of the soap film. From geometry, light rays reflected off the front and back of the soap film have a round-trip path difference of $2t/\cos\theta$, where t is the film thickness and θ is the angle of refraction—which is related to the viewing angle through Snell's law. The condition for constructive interference is that the path difference for these two light rays equals a half integral $(m + 1/2)$ multiple of the wavelength of light inside the film (λ/n), where n is the index of refraction. (The extra factor of 1/2 arises because of the 180-degree phase change when light is reflected off the front surface, which effectively reverses the condition for constructive and destructive interference.) Thus, the condition for constructive interference is

$$\frac{2t}{\cos\theta} = \left(m + \frac{1}{2}\right)\frac{\lambda}{n}.$$

Based on this equation, the observed color of the illuminated film (the wavelength that constructively interferes), will depend on the viewing angle. Notice that the color will also depend on the film thickness t, so that if you blow more air into the bubble or let it shrink, you can systematically alter its thickness and apparent color, while keeping the viewing angle constant.

Observers might be asked to predict whether the color change will be toward the red or blue end of the spectrum when the bubble is inflated. If the film appears yellow for example, inflating the bubble decreases its thickness, and therefore makes a shorter value of λ satisfy the preceding equation for the same value of the integer m, so its color should shift toward the blue end of the spectrum. But if the film already appears blue, it should cycle through the colors starting at the other (red) end of the spectrum in order to satisfy the preceding equation using the next lower value of the integer m.

After a while, you may find that the color of the bubble gradually shifts to yellow and eventually to white. The explanation for this fading of the colors has to do with the variation of bubble thickness from top to bottom, which gradually increases over time due to gravity. As a result of gravitational settling, the condition for constructive interference is satisfied by every color (at different heights on the bubble), and hence the image appears white, since it is formed from rays hitting the bubble surface at various heights. In other demos using *flat* soap bubbles on a circular wire frame, the effect of this variation in thickness is found to create a series of multicolored horizontal bands—see, for example, *Turning the World Inside Out.*[3]

11.4 Poisson's bright spot

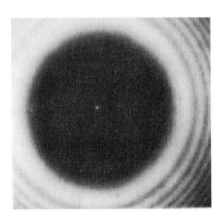

Demonstration
Poisson's bright spot in the middle of a circular shadow of a small sphere can be observed either by the naked eye (with no equipment whatsoever), and also by using the pinhole imaging system of demonstration 10.10.

Equipment
A pin with a spherical head, a small converging lens having a focal length about 1 cm, a lens holder, an unfrosted bulb, a cardboard carton about 0.5 meters on a side, and pieces of aluminum foil taped onto 35 mm slide frames.

Discussion
The diffraction pattern in the photograph with Poisson's spot in the center is courtesy of P. M. Rinard, from his article appearing in the American Journal of Physics.[7] The appearance of a bright spot in the middle of the circular shadow of a round object was first observed in 1723 by Giacomo Maraldi, but it was forgotten, and not explained until nearly a century later by Simon-Denis Poisson. Ironically, Poisson's prediction of a bright spot was made with the intent of *dis*proving the wave theory of light. Poisson, in attacking the wave theory put forward by Augustin Fresnel was in effect saying that the theory had to be wrong, because it predicted an absurdity—a bright spot in the middle of a shadow.

Interference and Diffraction

Here are two methods for observing Poisson's bright spot—also sometimes called Arago's or Fresnel's bright spot. The first method suggested to me by Dan Styer of Oberlin College is so simple that it requires no equipment at all. Look at a bright but diffuse light source, such as a fluorescent tube. Squint, and you will see various moving shapes floating across your field of view. These shapes, known as "floaters" are the diffraction patterns of small impurities or defects floating inside your eyeball. Some of the floaters will be small and circular, and they will all be seen to have a bright spot at their center—Poisson's spot! Of course, in order for this explanation to be convincing, we must take for granted that the floaters do not in fact have small holes at their center—otherwise the bright spot at their center might have nothing to do with diffraction!

Because of this uncertainty, it would be far preferable to look for Poisson's spot in the diffraction pattern from a small object known to have a spherical or circular shape lacking any holes. The second technique for observing Poisson's bright spot, described below, uses the pinhole imaging system of demonstration 10.10, and unfortunately allows only individual viewing. The method is based on that given by Nahum Kipnis.[6] Here are the important procedures you should observe according to Kipnis.

- Use a pinhole in aluminum foil taped against the box containing the unfrosted bulb. You may want to experiment with several size pinholes made with a sewing needle. Exact roundness is unimportant.
- It is important to place your eye along a line of sight from the pinhole, where the light from the filament has maximum brightness. Place the small lens along that line of sight, about a meter from the pinhole, and adjust the elevation of the lens above the table to correspond to the direction of greatest brightness. (It would be simplest if your lens holder had an elevation adjustment, but if not you could put it on books.)
- View the lens looking toward the pinhole from a distance of 10 to 20 cm from the lens, and you should see a bright circle covered with specks—which is the out of focus image of the illuminated spot on your retina.

- Adjust the lens to make the circle as bright as possible, and move your eye toward the lens until the circle fills your field of view.
- Place the pin with a round head between the lens and the pinhole about 10 to 20 cm in front of the lens. Position the pinhead carefully so that its circular shadow is in the center of the bright field of view.
- You should observe concentric colored rings inside the circular shadow of the pinhead, and possibly a bright spot at the center—Poisson's spot.
- If you don't see the bright spot at the center of the shadow, try varying the distance of the pin.

Notes

1. D. R. Carpenter Jr. and R. B. Minnix, *The Dick and Rae Physics Demo Notebook*, p. W-335.

2. A. L. Schawlow, The American Journal of Physics, **33**, 922 (1965).

3. R. Ehrlich, *Turning the World Inside Out*, p. 205.

4. Actually, according to Ron Edge in an informal communication, a soap bubble under a bell jar or a two-liter plastic bottle with its bottom cut off will have a very long life.

5. F. Behroozi and D. W. Olson, The American Journal of Physics, **62**, 856–57 (1994).

6. N. Kipnis, *Rediscovering Optics*, Minneapolis, MN: BENA Press, 1993, and also a workshop he conducted at the 1995 AAPT Summer meeting in Spokane, WA.

7. P. M. Rinard, The American Journal of Physics, **44**, 70 (1976).

Chapter 12

Modern Physics

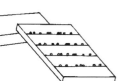

12.1 Chain reaction simulation

Demonstration
The concept of a chain reaction can be illustrated using an inclined board with rows of marbles, in which an avalanche is started with a single marble.

Equipment
A book, board, or piece of lucite (if doing the demonstration on the OHP), some rubber bands, and a collection of marbles.

Discussion
Stretch four rubber bands around the book or board, so as to make equally spaced parallel bands. Prop up one end of the board parallel to the bands, so the board makes a small angle with the horizontal. The rubber bands serve as shallow ledges on which you can place marbles without them rolling down the board. Call the maximum number of marbles that can fit on a ledge, N_{max}. Place an equal number of marbles on each ledge, N, at random locations. Now place a single marble above the first ledge and let it roll. If N is an appreciable fraction of N_{max}, then one marble will create a chain reaction, or an avalanche, in which an ever-growing number of marbles leave each ledge due to collisions.

The ledges in this simulation correspond to the individual generations in a nuclear fission process. The rapidity with which the number of rolling marbles increases clearly depends on the average density of marbles on each ledge. If, for example, the average number of marbles struck by one incoming marble equals one, then the number of marbles taking part in "the reaction" doubles each generation—at least initially. Unfortunately, the simulation does not faithfully convey the concept of critical mass, however, because the number of marbles at each generation is always at least as great as the number at the preceding generation unless an appreciable fraction of marbles roll off the edge of the board.

An alternate simulation of a chain reaction suggested by Robert Marzewski, *does* allow the ideas of critical mass and critical density to be demonstrated.[1] The demonstration also involves the active participation of an entire class. You first need to assemble some number of students in a group at the center of the room. Give each student two corks or two crumpled pieces of paper, which they are instructed to throw high in the air if and when they feel a cork or piece of crumpled paper hit them.

Have all the students close their eyes, and start the "reaction" by tossing one cork into the group. If the number of students (the mass) is small, there is a good chance a chain reaction will not occur, because of the probability that both corks thrown by the first student will leave the boundaries of the group and not hit any other students. For a certain number of students (the critical mass), the probability is that, on the average, one of the two corks thrown by each student stays within the group, and hits other students. For an appreciably larger number of students, a runaway chain reaction will occur, with nearly everyone eventually throwing corks. The simulation also illustrates the concept of critical density, because a critical mass for one student spacing becomes subcritical if the students increase their spacing, causing an increase in the probability of a thrown cork missing everyone.

12.2 Nuclear fusion simulation

Demonstration
A nuclear fusion reaction can be simulated using two magnetic marbles on the OHP.

Equipment
A transparent, grooved plastic ruler, two magnetic marbles, and a piece of folded index card.

Discussion
A key aspect of the nuclear fusion reaction is that high temperatures are needed for its initiation, in

order to overcome the repulsive Coulomb barrier between the colliding particles. This aspect of nuclear fusion can be simulated using magnetic marbles. Place the two marbles some distance apart in the groove of a plastic ruler, and put a piece of folded index card between them. When you roll the marbles toward each other forcefully they will fold the card and stick together with the folded card between them—they have overcome the simulated Coulomb barrier owing to their high initial velocities. However, if they are rolled toward each other less forcefully, they should bounce off the springlike card and recoil.

To achieve success with this demo it is quite important to use the full length of an index card when cutting off a piece. Using a long piece of index card ensures that: (1) repulsion begins when the balls are relatively far apart, and (2) the repulsive force that the card exerts is nearly along the direction of the ruler groove, and therefore will be unlikely to knock the marbles out of the groove. You may also need to experiment with the placement of the folded card and the opening angle of the fold.

The reason that this demo faithfully simulates the force between colliding light nuclei is that the attractive force between the marbles has a dependence on distance that assures it dominates over the repulsion at short (but not long) distances. We may think of the magnets as approximately magnetic dipoles, which have an attraction that varies as the inverse fourth power of their separation. In contrast, the repulsive force created by the folded index card does not increase as fast as the marble's separation decreases.

Although this demonstration simulates one aspect of the fusion reaction—overcoming the Coulomb barrier with sufficiently high collision speeds—it is less faithful in several other key respects. Unlike the actual exothermic fusion reaction, where energy is liberated in the form of kinetic energy of recoiling particles, in the simulation no particles are emitted when the magnetic marbles join. In addition, in the actual case of fusion the reaction can proceed even though the colliding particles do not have enough energy on a classical basis, due to the process of quantum mechanical tunneling.

12.3 Quantum mechanical tunneling analogy

Demonstration
The process of "frustrated total internal reflection"—an analog to quantum mechanical tunneling can be demonstrated using a glass of water.

Equipment
A glass or transparent plastic cup of water.

Discussion
When pressing your fingertips hard against the sides of the glass you should be able to see the ridges making up your fingerprints, when looking down into the water. But if you hold the glass more loosely, you will not see the ridges against the sides of the glass. This observation demonstrates the process of "frustrated total internal reflection." Mathematically, the basis of total internal reflection is that a light wave incident on a boundary between two media at an angle greater than the critical angle can exist in the lower index medium only in the form of a decaying exponential function. Normally, the transmitted decaying exponential wave rapidly goes to zero, so 100 percent of the wave must be reflected when light is incident past the critical angle.

However, if the second medium only extends for a short distance, the transmitted wave can make it through this "forbidden" region because the exponential wave form does not decay to zero, but joins smoothly on to a sinusoidally varying wave on the other side. The transmission probability of the light wave through a barrier is therefore a negative exponential function of the width of the barrier—exactly like it is in the case of quantum mechanical tunneling of electrons through barriers.

So, why do your fingerprints appear when you press hard on a glass? When you press your fingers firmly on the sides of the glass, an appreciable layer of air between your fingers and the glass exists only in between the ridges of your fingerprints. For the ridges themselves the air gap between them and the glass is reduced enough so that the light can tunnel through, and you see no reflection along the ridges.

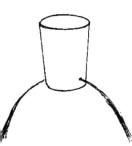

12.4 Weightlessness

Demonstration
The unobservability of gravitational effects in a freely falling reference frame can be illustrated by dropping a water-filled cup with two holes in it.

Equipment
A styrofoam cup.

Discussion
As suggested in *The Dick and Rae Physics Demo Notebook,* you can easily demonstrate weightlessness using a water-filled styrofoam cup.[2] Make two holes in the sides of the cup near the bottom. Cover the holes with your fingers, and then uncover them just before you drop the cup from a height of about 2 meters. During the brief free fall you should observe that no water flows out of the holes. At one level, this demonstration is "simply" an illustration of the concept of weightlessness in freely falling reference frames. At another level, it illustrates the concept underlying Einstein's Equivalence Principle—which states that locally gravity is equivalent to being in an accelerated reference frame.

12.5 Tippy tops and spinning electrons

Demonstration
The tippy top toy can "illustrate" the two states of electron spin.

Equipment
A tippy top toy consisting of a spherical hemisphere with a stem used to spin it by hand.

Discussion
When given a spin on its spherical side, the tippy top turns over and spins on the stem. If this demonstration is used during discussions of the "up" and "down" electron spin states, it should be shown more as a joke than a real illustration of a transition between two spin states. First, because the quantum mechanical situation really has no classical analog.

And second, because the top spins in the *same* direction before and after the flip, so it is not really correct to view the transition as one from spin "up" to spin "down." The top's inversion is an interesting phenomenon in its own right, apart from any analogy with spinning electrons. The flip occurs as a result of a frictional torque at the point of contact, so it should take longer to occur if the top is spun on a very smooth surface.

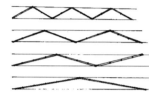

12.6 Time dilation simulation

Demonstration
The effects of time dilation in a "bouncing light pulse clock" can be illustrated using a wooden folding 6-foot-long ruler.

Equipment
A 6-foot-long folding wooden ruler.

Discussion
One common way of deriving the time dilation formula uses a clock in which a pulse of light bounces up and down between a pair of parallel mirrors—with one tick of the clock defined by a round-trip of the light pulse. An observer watching the clock move past along a direction parallel to the mirrors obviously would see the light pulse travel along a diagonal zigzag path in order to remain inside the clock. The constancy of the speed of light for all observers requires that in a given proper time corresponding to one tick, all observers agree on the total length of path of the light pulse. As a result, the number of bounces the light pulse makes is less for an observer watching the clock go past than for someone seeing the clock at rest—with the effect being more pronounced the faster the speed of the clock.

A 12-segment folding wooden ruler can be used to represent the zigzag path of a light pulse for clocks seen moving past at various speeds. For example, when completely folded up and held vertically it represents 12 bounces inside a clock that is at rest relative to the observer. If you unfold the ruler to make a zigzag path where each diagonal element consists of two segments of the ruler, and the height of the zigzag remains the same as it was in the folded ruler (0.5 feet), this represents a clock ticking 6 times instead of 12 (time running slow by a factor of two—see top part of figure. What factor slow down in time do the other three parts of the drawing represent?)

You can measure the speed of the clock as a fraction of the speed of light by measuring the horizontal distance *in feet* between successive zags. It is important to measure the distance in feet, because in a clock at rest, the bouncing light pulse would travel one foot for a round-trip, so in our system of units a light pulse travels one foot in one tick of a stationary clock. You can also arrange the 12-segment ruler to correspond to the path taken by a light pulse clock in which time slows by a factor N, where $N = 3, 4, 6,$ and 12. (In the $N = 6$ case, for example, the ruler needs to have only one bend, and its midpoint should be 0.5 feet higher than either end—see bottom part of figure.) In general, you will find (based on the Pythagorean theorem), that the speed of the clock for which time runs slow by a factor of N is given by

$$v/c = \sqrt{1 - 1/N^2}.$$

12.7 The "tachyon telephone"

Demonstration
The concept of *hypothetical* faster-than-light tachyons, and their potential ability to send messages back in time can be illustrated using transparencies on the OHP.

Equipment
Several transparencies, and a piece of opaque cardboard into which you have cut a narrow slit. (You may find it helpful to use some tape to smooth the edges of the slit.)

Discussion
Worldlines on a two-dimensional space-time diagram show the complete history of an object. Using the appropriate units (seconds and light seconds, for example), and plotting time on the vertical axis, a light beam can be shown using a 45-degree line, while an object traveling faster than the speed of light—such

as the hypothetical tachyon—will have a slope less than 45 degrees. To illustrate the strange behavior of tachyons, draw two space-time diagrams for the process in which a pair of approaching masses exchange a particle between them, and as a result of the exchange, the masses separate. Let the exchanged mass leave the left mass, and be received by the right one. In the first version of the diagram (figure on left), make the exchanged particle have a slope of greater than 45 degrees, and in the second version (figure on right), make its slope less than 45 degrees—corresponding to an exchanged tachyon.

To view the unfolding "real time" evolution of the exchange processes represented by these two space-time diagrams, we need to view them at a succession of times through a horizontal "time window" or slit that can be slowly moved upward along the vertical time axis. What appears momentarily in the time window shows the moment by moment sequence of events. For both the exchange processes, the sequence of events will appear the same as you slide the time window upward: (1) the two masses approach each other, (2) an exchanged object leaves (is transmitted by) the left mass, (3) it is received by the right mass, and (4) the masses recede. Now let us view each process in a different reference frame. In a reference frame moving with respect to the first one, the time window will no longer be horizontal as time advances, but it will make a slope with the horizontal—a slope which depends on the speed of the relatively moving observer.

Orient the time window along a slope that exceeds that of the tachyon world line, and move it onward (upward) in time. As the time window advances, the sequence of events is the same as before for the case when a non-tachyon is exchanged, but with the important difference that now steps (2) and (3) are reversed in their time sequence for the tachyon exchange. Effectively, the tachyon is received before it is emitted, or it was sent back in time! Before you get too upset by this idea, remember that no tachyons have ever been observed, and the possibility of using them to send messages backward in time makes many scientists think they never will be observed.

But, there are alternative interpretations that could avoid backward time transmissions using tachyons, and might actually allow them to exist. In one such alternative interpretation, the identity of the sender and receiver of messages get switched when the exchange is observed according to some moving observers. Of course, this possibility itself raises some very troubling issues, because it implies that either the distinction between cause (transmission) and effect (reception) is an illusion, or else that the emission/reception of tachyons are uncontrollable processes of no use in transmitting information.

12.8 Michelson-Morley experiment simulation

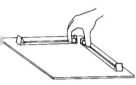

Demonstration
The basic idea of the Michelson-Morley experiment can be illustrated using rolling balls on the OHP.

Equipment
Two 1-inch (2.54 cm)-diameter stainless steel balls, some index cards, and a piece of 20 x 20 centimeter piece of thick cardboard to construct the apparatus.

Discussion
The crucial observation in the famous Michelson-Morley experiment was that light seemed to take the exact same amount of time to travel in the two arms of the apparatus, no matter what its orientation with respect to a hypothetical "ether wind"—whereas a calculation based on classical physics would predict there should be a difference depending on the orientation of the apparatus. The basis of this expectation (*not* borne out in the actual experiment), can be explained using an apparatus in which rolling balls are used to represent light pulses traveling back and forth between mirrors.

   The apparatus consists of two equal-length channels in which the two balls can roll equal distances along paths at right angles to one another. The two one-centimeter-wide channels may be cut out of the 20 × 20 centimeter piece of cardboard, so that balls rolling in them will be visible on the OHP. In addition, folded pieces of index card (serving as gentle springs), need to be taped to each end of the two channels. Initially, the balls are placed at the beginning of each channel, and held in place by your thumb and index finger, so as to compress the pieces of folded index card.

By simultaneously removing your fingers from each ball, and allowing the index cards to unfold, you launch them with roughly identical velocities down the two right-angle channels. Pieces of index card at the far end of each channel act like springs to reflect each ball back to its starting point. If the pieces of index card have the same amount of springiness, each ball should complete its round-trip in the same time. If that is not the case, try adjusting the amount of fold in the index card pieces.

At this point, we need to introduce the equivalent of an "ether wind" into the demonstration. This can be accomplished one of several ways. One way would be to drag the apparatus across a table top at constant speed as the balls are launched, which simulates the process of the apparatus moving through a stationary ether. But, of course, the same results can be obtained if the apparatus is fixed while an ether wind blows past—which is more suitable for showing the demo on the OHP. In this case, you need to keep the apparatus fixed with a transparency underneath it, and drag the transparency as the balls are launched. (Remember that the channels have no bottoms, so the balls roll on the moving transparency.) You want to pull the transparency at a slower speed than that of the balls, otherwise the ball which rolls upstream (against the ether wind) will never make it back to its starting point. The transparency must be long enough, so that it remains under the device for an entire round-trip of the balls. Be sure to draw some waves on the transparency, so that it will be visible as it is pulled.

Which ball should complete a round-trip faster when an ether wind is blowing? In the classical calculation based on addition of velocities, the velocity of the ball in the "downwind" direction is $c + v$, while the velocity in the "upwind" direction is $c - v$, where c is the speed of the ball and v is the speed of the ether wind. Therefore, the time for a round-trip for the ball traveling downstream and then upstream a distance L would be predicted to be $t_1 = L/(c + v) + L/(c - v) = 2Lc/(c^2 - v^2)$. For the ball traveling cross-stream and back, the predicted round-trip time from the classical calculation is $t_2 = 2Lc/\sqrt{c^2 - v^2}$, where $t_2 < t_1$. Thus, you might expect that the ball that makes the cross-stream trip completes its round-trip sooner than the one that goes up and downstream.

However, when you try the demonstration with an "ether wind," you will likely find that the reverse is true—it is the "crosswind" ball that has the greater delay in its round-trip. The anomaly is explained by the effects of rolling motion. The crosswind balls develop a pronounced spin along an axis pointing in the direction of the channel which greatly impedes their ability to roll in that direction. (Presumably, if we did the demo using low friction sliding pucks, this defect would be remedied.)

Nevertheless, despite this departure from the classical prediction, the demo is still useful, because it faithfully reproduces the essence of the expectation Michelson and Morley had of their experiment, namely, a *difference* in time in round-trip times in the two arms of their apparatus—a difference that depended on the orientation with respect to the ether wind. For example, it is most instructive to rotate the apparatus, so that the ether wind transparency is pulled at a 45-degree angle with the two channels. Due to symmetry, the effect of the ether wind should now be identical on the two balls, which should arrive back at their starting points simultaneously, assuming they started simultaneously.

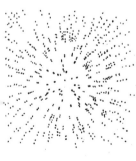

12.9 Expansion of the universe

Demonstration
A simulation of the expansion of the universe can be given on the OHP using two transparencies of random dot patterns.

Equipment
A random dot pattern.

Discussion
You need to make one or more random dot patterns which simulate a photograph of a collection of galaxies. You could make the pattern by photocopying those appearing in the article by David Chandler, in which this demonstration was described.[3] Alternatively, you could make a random dot pattern by exposing a piece of paper to a short burst from a spray

can of dark paint—or simply depositing randomly located dots by hand on a piece of paper. Either way, after you have a random dot pattern, photocopy two transparencies from it, making one about 7 percent enlarged. These two transparencies represent simulated views of the universe taken one billion years apart. Each dot on a transparency represents a galaxy or a cluster of galaxies.

Superimpose the two transparencies on the OHP, keeping their orientations the same, and observe how the expansion appears to be centered on one particular point (where a particular galaxy occupies the same position on both transparencies—the central point in the figure accompanying this demo). Now slide one transparency over the other, and observe how the center of the expansion shifts to a new point—thereby illustrating how, in a universal expansion, an observer on any one galaxy sees all the others receding.

You can use the dot patterns to deduce how much the simulated universe expanded in one billion years by measuring the percentage increase in distance from the center of the pattern for a sample of easily recognized dots—it should of course be 7 percent, if that was the extra magnification you used to make one of the transparencies. The age of the universe, assuming a uniform expansion without deceleration would, in this case, be 1 billion years divided by 0.07, or around 14 billion years. (If only measuring the age of the real universe were that easy!)

This simulation lacks realism in a number of important respects:

- It ignores the random motions of galaxies and galactic clusters that is superimposed on the universal expansion. These appreciable random motions introduce a considerable uncertainty in the exact value of the expansion parameter (the Hubble Constant).
- It ignores systematic departures from a universal expansion such as would occur near the "Great Attractor." In the vicinity of such an object, all galaxies show a motion toward this body superimposed on the universal expansion. Including this bit of realism in the simulation would not be easy, because you would need to modulate the magnification near a certain point in a position-dependent way.

- The size of galaxies does not enlarge as their separation increases—an error not made if you use the pair of images in Chandler's article, instead of two photocopies of a paint spatter. (But, in any case, even with the paint spatter pattern, the size of the galactic inflation may be too small to notice.)

12.10 Pole in the garage paradox

Demonstration

The pole in the garage paradox of special relativity can be presented with the aid of a collapsible pointer.

Equipment

A collapsible pointer and two cardboard boxes. One of the boxes should be half as deep or high as the other. The height or depth of both boxes should both exceed the length of the collapsed pointer, but by less than a factor of two in the case of the smaller box.

Discussion

Hold a fully extended pointer at rest, illustrating its "proper length." Now move the pointer past, and collapse it by a greater and greater amount as its speed increases. Be sure to explain that in the actual case of length contraction no physical force is responsible for the measured shrinkage in length. Move the pole (contracted to half its full length) toward the larger open box (the "garage"), which is deep enough to contain the pole only when it is fully contracted. Observe that the door to the garage can be closed, thereby trapping the contracted pole inside. (We are assuming in this simulation that the garage rear wall is made of an impenetrable material.)

The basic paradox is that when we view the previous scene from the point of view of someone carrying the moving pole, it is the garage not the pole that is contracted. To visualize this situation, extend the pointer so that it is twice its collapsed length (which we take to be its proper length). Also, use the box which is half as deep as the original box in this case, and move it toward the stationary pole. Apparently, there is no way for the pole to be contained within the garage in this reference frame. However, relativity comes to the rescue here, because one of its tenets is that no signal can travel faster than the speed of light, which means there can be no such thing as a rigid object. When the front of the stationary pole makes contact with the back of the onrushing garage, the back of the pole must remain stationary (and hence the pole collapses). This collapse must occur, because no signal can travel faster than the speed of light, so that it takes a finite time (at least L/c) for the back of the pole to learn that its front has made contact and started moving. The pole does fit in the garage after all.

This paradox is sometimes explained on a slightly different basis using the relativity of simultaneity. By the phrase "pole inside the garage," we mean that the two ends of the pole are simultaneously within the coordinates defining the front and back of the garage. The two observers disagree about whether this can occur while the pole is in motion, because their judgments over simultaneity differ. Nevertheless, both observers agree on the final outcome: a crumpled up pole found inside the garage.

Notes

1. R. Marzewski, The Physics Teacher, November 1988.
2. D. R. Carpenter Jr. and R. B. Minnix, *The Dick and Rae Physics Demo Notebook*, p. S-055.
3. D. Chandler, The Physics Teacher, February 1991.

Bibliography

A good summary of the literature on physics demonstrations up to 1979 can be found in John A. Davis and Bruce G. Eaton, "Resource Letter PhD-1: Physics Demonstrations," American Journal of Physics 47(1), 835–40 (October 1979). Some well-known books are listed below.

Blasi, Rocco C., ed., *Physics Fun and Demonstrations with Professor Julius Sumner Miller*. Franklin Park, IL: Central Scientific Company, 1974.

Bohren, C. F., *Clouds in a Glass of Beer.* New York, NY: John Wiley and Sons, 1987.

Carpenter, D. Rae Jr., and R. B. Minnix, *The Dick and Rae Physics Demo Handbook*. Lexington, VA: Dick and Rae, Inc., 1993.

Doherty, P., and Rathjen, D., eds., *The Exploratorium Science Snackbook.* San Francisco, CA, 1991.

Edge, R. D., *String and Sticky Tape Experiments.* College Park, MD: American Association of Physics Teachers, 1981.

Ehrlich, R., *Turning the World Inside Out and 174 Other Simple Physics Demonstrations*. Princeton, NJ: Princeton University Press, 1990.

Freier, G. D., and F. J. Anderson. *A Demonstration Handbook for Physics*. 2d ed., College Park, MD: American Association of Physics Teachers, 1980.

Jones, E. G., *Physics Demonstrations and Experiments for High School*. Physics Department, Mississippi State University, 1984.

Liem, T. L., *Invitation to Science Inquiry*. 2d ed., Chino Hills, CA: Science Inquiry Enterprises, 1987.

Meiners, H., ed., *Physics Demonstration Experiments, volumes 1 and 2.* New York: Ronald Press, 1970.

Morse, R., *Teaching About Electrostatics.* College Park, MD: American Association of Physics Teachers, 1992.

Sutton, R., *Demonstration Experiments in Physics.* New York: McGraw-Hill, 1938 (now available as a reprint from AAPT).

Taylor, C., *The Art and Science of Lecture Demonstrations.* Philadelphia, PA: Adam Hilger, 1988.

VanCleave, Janice Pratt, *Teaching the Fun of Physics.* New York: Prentice Hall, 1985.

Bibliography

Walker, J., Carroll, B., Davis, J. A., and Berg, R., *The Video Encyclopedia of Physics Demonstrations*. The Education Group, 1992.

Walker, Jearl, *The Flying Circus of Physics*. New York: John Wiley & Sons, 1975.

Index